Contents

An Introduction to Water Quality Modelling

Second Edition

Edited by

A. James
Department of Civil Engineering
University of Newcastle upon Tyne, UK

JOHN WILEY & SONS

Chichester · New York · Brisbane · Toronto · Singapore

Copyright © 1993 by John Wiley & Sons Ltd,
Baffins Lane, Chichester,
West Sussex PO19 1UD, England

Other Wiley Editorial Offices

John Wiley & Sons, Inc., 605 Third Avenue,
New York, NY 10158-0012, USA

Jacaranda Wiley Ltd, G.P.O. Box 859, Brisbane,
Queensland 4001, Australia

John Wiley & Sons (Canada) Ltd, 22 Worcester Road,
Rexdale, Ontario M9W 1L1, Canada

John Wiley & Sons (SEA) Pte Ltd, 37 Jalan Pemimpin #05-04,
Block B, Union Industrial Building, Singapore 2057

Library of Congress Cataloging-in-Publication Data

An Introduction to water quality modelling / edited by A. James. —
    2nd ed.
        p.   cm.
    Includes bibliographical references and index.
    ISBN 0-471-92347-8
    1. Water quality—Mathematical models.   2. Water quality
    management—Mathematical models.   I. James, A.
    TD370.I575   1993   92-14426
    628.1'61—dc20   CIP

British Library Cataloguing in Publication Data

A catalogue record for this book is available from the British Library

ISBN 0-471-92347-8

Typeset in 10/12pt Times by QBF, Salisbury, Wiltshire
Printed and bound in Great Britain by Bookcraft (Bath) Ltd

List of Contributors

P. M. BERTHOUEX   Department of Environmental Engineering
University of Wisconsin, USA

D. J. ELLIOTT   Department of Civil Engineering
University of Newcastle upon Tyne, UK

A. JAMES   Department of Civil Engineering
University of Newcastle upon Tyne, UK

R. MACKAY   Department of Civil Engineering
University of Newcastle upon Tyne, UK

S. MARKHAM   Department of Civil Engineering
University of Newcastle upon Tyne, UK

M. S. RILEY   Department of Civil Engineering
University of Newcastle upon Tyne, UK

Preface to First Edition

In the last two decades the application of modelling techniques to water quality problems has increased dramatically and many useful techniques have emerged. Unfortunately, teaching of these techniques has not received the same attention and has consequently lagged behind. The hiatus is apparent in the large number of papers and books published for the advanced practitioner compared with the dearth of simple texts for the beginner.

This book is aimed at the beginner. It is based upon a number of courses which have been held in the Civil Engineering Department of the University of Newcastle upon Tyne to introduce quality modelling to scientists and engineers who were working, or intending to work, in water pollution control.

The text is a simple guide, suitable for biologists, chemists, engineers and others without any previous knowledge of modelling or computing. The level of mathematical complexity has been deliberately restricted.

An introductory chapter explains the concepts and terminology used in simulation, optimization and computer-aided design. This is followed by two other introductory chapters dealing with computing and numerical methods. The chapter on computing includes a guide to BASIC which is the simplest programming language in common use and the language used throughout the remainder of the book.

The final introductory chapter explains the hydraulic, chemical and biological ideas which are used in formulating models of water quality.

The remainder of the book is divided into two sections dealing with the application of modelling techniques to water pollution and wastewater treatment plants.

Each chapter presents the modelling concepts and shows how these may be built into models suitable for examining frequently occurring problems. A complete listing of an example program is included as an appendix to each chapter.

It is hoped that this book will help and encourage students and practitioners in the water quality field to understand mathematical modelling techniques.

Preface to Second Edition

In the eight years since the publication of the first edition, there has been a tremendous increase in the use of mathematical models in environmental engineering, especially for the control of pollution in rivers and estuaries. Modelling has also addressed a much wider range of pollutants and there has been an increase in the range of conceptual approaches to the formulation of models.

The text of this book has therefore been modified to reflect these changes. The chapters dealing with techniques have been expanded to cover a greater range of kinetics and introduce a background of understanding for statistical techniques and time series analysis.

Similarly the chapters dealing with the applications of models to rivers, estuaries, lakes, groundwater and the marine environment have been expanded and updated.

Unfortunately this has resulted in the omission of the section dealing with the modelling of wastewater treatment, but it is hoped that this can be covered by a separate volume.

The overall aims of the book have not changed and this remains an introductory text for people wishing to learn about water quality modelling.

Chapter 1

Simulation

A. JAMES

1.1 INTRODUCTION

The term simulation has been used by a variety of authors to describe varying classes of models. Simulation is used here in its widest and most basic sense, to describe all types of models which represent water quality changes in some mathematical form. This definition therefore includes all types of mechanistic models in which all processes are deterministically represented and also statistical models. Not included in this definition are models intended to optimize rather than simulate.

1.2 SIMULATION TECHNIQUE

The creation of a mathematical model, which simulates changes in water quality, can be carried out in a wide variety of ways, but it always involves the following two stages:

(1) Analysis of the system—this may be theoretical, observational, or experimental.
(2) Synthesis of a mathematical replica of the system.

The aim in formulating the simulation model should be to achieve maximum simplicity consistent with the required degree of accuracy and detail. The steps in the formulation process are summarized in Figure 1.1 and may be explained as follows:

(a) *Formulation of objectives.* This must be done as clearly and as quantitatively as possible. Failure to clarify the objectives usually leads to the production of an

An Introduction to Water Quality Modelling. 2nd Edition. Edited by A. James
© 1993 John Wiley & Sons Ltd.

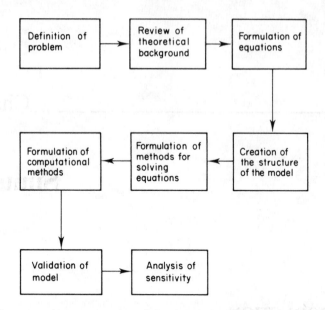

Figure 1.1. Diagrammatic representation of the steps in creating a simulation

unsatisfactory model. On the other hand, a clear quantitative statement of the output required considerably simplifies the subsequent steps in formulation. Care is therefore required in adapting existing models to new purposes in case the objectives are incompatible with the model structure.

(b) *Review of theoretical background.* This is a desk study of any previous attempts at formulating models for the same or similar situations. It should include an analysis of the processes affecting water quality, methods of representation, and appropriate ranges for any model parameters.

(c) *Formulation of the model.* This involves a decision about the type of model (i.e. deterministic, steady-state, etc.) and which processes to include. Care should be taken to avoid over-complication of the model, by eliminating relationships that do not significantly affect the output. It is worth while to examine alternative types of models before making a final choice. The implications for data collection should be carefully noted. A conceptual diagram is often useful at this stage. Figure 1.2 is an example.

(d) *Creation of the model structure.* This is a difficult stage of model building, particularly with complex models. It is generally advisable to begin by identifying the large subdivisions of a model and to proceed by fitting these together in diagrammatic form with a flow chart. This helps establish the overall flow of information through the model and enables individual subroutines to be developed and tested separately. It also helps to make models more versatile. The example of the modular structure shown in Figure 1.3 could easily be adapted to handle other water quality parameters by inserting appropriate subroutines for sources and sinks.

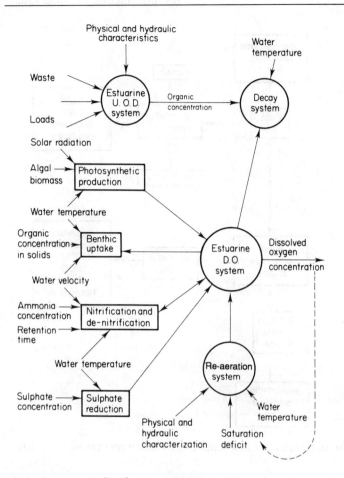

Figure 1.2. Dissolved oxygen regime in an estuary

(e) *Formulation of equation.* Based upon the review of theory and the model structure it should be possible to state the relationships involved in some formal mathematical or statistical way. Adoption of a hierarchical approach to this process often results in a clearer set of equations in which the influence of primary and secondary relationships can be more easily appreciated. In common processes like dispersion, the main problem may be to choose the best equation from several formulations. Some preliminary data may be needed to guide the choice. Once the equations have been established, it is possible to describe the information flow within each subroutine or subdivision of the model. Analogue diagrams have sometimes been used for detailed representation of information flow (see James 1984), but this use has generally been abandoned.

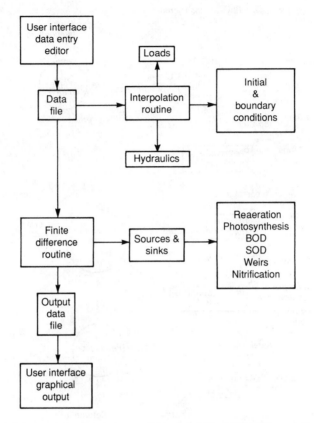

Figure 1.3. Modular structure of a water quality model (the DYNUT model)

(f) *Formulation of methods of solution.* In a few special cases it may be possible to solve the equations analytically, but most models involve the use of numerical methods for solving partial differential equations, interpolation, etc. The choice of the appropriate numerical technique is crucial for numerical stability and accuracy and also for minimizing computational effort (although this latter aspect has become less important). Numerical techniques are discussed in Chapter 3.

(g) *Production of computer programs.* The computation step may be carried out on any suitable machine, from a calculator to a mainframe computer. For more complex models, computers are essential; even simple models may require computers where many runs are required. The form of input and output, and the choice of language, are largely dictated by the facilities available. Special modelling languages like CSMP (continuous system modelling programs) have been devised but are not as widely used as programming languages like Pascal, C or FORTRAN. The simplest programming language is BASIC, which, together with TURBOBASIC and QUICK-BASIC, has provided versatile and powerful tools for many modelling purposes. Computer programming is discussed in detail in Chapter 2.

(h) *Calibration and validation.* Validation is an obvious requirement since no model may be accepted as representing a system without suitable proof. The first stage is to calibrate the model, i.e. values of parameters in the model are adjusted so that the output from the model is re-run using other sets of input data and the output is compared with the observed output. This process may need to be repeated before the model gives results within an acceptable level of error. It is essential that completely independent data sets are used for calibration and verification.

(i) *Sensitivity analysis.* This is carried out to determine the sensitivity of the model output to changes in the values of the parameters in the model. Relative sensitivity can be used in allocating the time and effort to evaluate their numerical value.

It must be understood that the formulation of a water quality model is not achieved by a single pass through these steps. Decisions made at later steps in the chain may require earlier steps to be reconsidered, hence the two-way arrows in Figure 1.1.

1.3 SIMULATION EXAMPLE

An example of a simulation problem is illustrated in Figure 1.4. This shows a proposal to discharge the effluent from a coke-oven works into a stream some distance upstream of an abstraction point for a municipal water treatment plant. The aim of the model is to provide guidance in setting consent conditions (discharge standards) to be imposed on the discharge. The steps in the formulation of the model are summarized below.

(a) *Formulation of objectives.* These need to be stated as precisely as possible, preferably in quantitative terms. The potential impact of coke-oven effluent can be appreciated by examining the data in Table 1.1. This shows that although there are four significant pollutants, phenol is the most likely to cause problems.

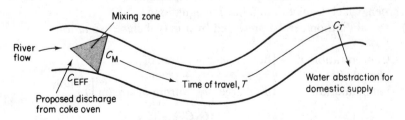

Figure 1.4. Example of a water quality simulation problem

Table 1.1. Background Information required for setting discharge standards

Pollutant	Typical concentration in coke-oven wastewater	Concentration causing problem in water treatment
Ammonia	100–200 mg/l	> 1 mg/l
Cyanide	10–50 mg/l	> 1 mg/l
Phenol	100–200 mg/l	> 0.001 mg/l
Thiocyanate	50–100 mg/l	> 5 mg/l

On the basis of this information the objective of the model becomes to determine the relationship between phenol discharged into the stream and the phenol concentration at the point of abstraction for a variety of environmental conditions (e.g. river flow and temperature).

(b) *Review of theory*. The concentration of phenol at the point of abstraction may be influenced by various factors, notably:

   (i)   initial dilution by freshwater flow;
  (ii)   losses due to volatilization;
 (iii)   dispersion due to molecular or eddy diffusion;
 (iv)   breakdown of phenol by bacterial activity;
  (v)   retention time (controlled by freshwater flow);
 (vi)   temperature effect on the breakdown rate.

At this stage it is useful to consider whether to represent all these processes in the model. In this example it has been decided to simplify the model by omitting dispersion and volatilization and, as a consequence, the calculated concentration will be slightly too high, or these will be included in the decay process. Other important decisions are with respect to variation in time and space. For simplicity, these have been reduced to one dimension (longitudinal) and time invariant (steady-state). Like all modelling assumptions these should be clearly stated and their implications must be appreciated.

(c) *Formulation of equations*. Following from the decisions made in (b) the processes chosen for inclusion need to be expressed in some mathematical or statistical format. For this example a simple, explicit, deterministic representation has been chosen, which leads to the following equations:

   (i)   Initial dilution can be presented by a mass balance equation:

$$\begin{array}{l}\text{Concentration of} \\ \text{phenol after} \\ \text{initial dilution}\end{array} = \left\{\begin{array}{l}\text{flux of phenol} \\ \text{from} \\ \text{discharge}\end{array} + \begin{array}{l}\text{flux of} \\ \text{phenol} \\ \text{from upstream}\end{array}\right\}\bigg/\begin{array}{l}\text{combined} \\ \text{flow}\end{array}$$

$$C_3 = \frac{(Q_1 C_1) + (Q_2 C_2)}{(Q_1 + Q_2)}$$

(ii) Decay of phenol in the stream could be presented as depending on the remaining concentration of phenol (i.e. a first-order reaction: see Chapter 5).

Concentration of phenol after decay $=\left\{\begin{array}{l}\text{Concentration of}\\\text{phenol after}\\\text{initial dilution}\end{array}\right\} \times \left\{\begin{array}{l}\text{Negative}\\\text{exponential}\\\text{decay coefficient}\end{array}\right\} \times$ retention time

$$C_4 = C_3 \exp(-K * RT)$$

(iii) Retention time could be represented by some function of the freshwater flow (assuming that the effluent flow is relatively small). The relationship would be established statistically from observed data shown in Figure 1.5.

Retention time = (max. retention time) − (coefficient) × (freshwater exponent flow)

$$RT = a - bQ_2^c$$

where Q_1 = effluent flow
Q_2 = upstream freshwater flow
C_1 = phenol concentration in effluent
C_2 = phenol concentration in upstream freshwater flow
C_3 = phenol concentration after initial dilution
C_4 = phenol concentration at abstraction point
RT = retention time
a, b and c are parameters from Figure 1.5.

In formulating these equations it is important to be aware of the assumptions that they contain. For example the initial dilution is assumed to take place

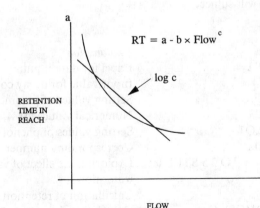

Figure 1.5. Relationship between retention time and flow

instantaneously and completely. These assumptions are valid only if (a) there is no significant density difference between effluent and upstream flow (for example, due to temperature or dissolved salts) and (b) the length of the mixing zone is short compared to the length of the river between the discharge and abstraction points. The length of the mixing zone depends on various factors, but for most bankside discharges it is around 50–100 times the width of the river.

Similarly, the neglect of dispersion and volatilization would need to be justified by theoretical evidence or experimental data.

(d) *Creation of the model structure.* In a model involving so few relationships, this is a trival exercise. The model structure can be represented by a flow diagram as shown in Figure 1.6. The only decision to be made is the range of environmental conditions that need to be explored. As with more complex models, the format of the input and the output should be considered. Menu-driven models with sophisticated graphical output have become popular but there is a trade-off for this ease of use in increased size, decreased portability (chiefly due to graphics), and increased complexity of programming, making it difficult or impossible for the user to modify this type of model.

(e) *Formulation of the method of solution.* For this method, this also is trivial since the only differential equation is

$$dC_3/dt = KC_3$$

which can be solved analytically.

If dispersion were shown to be significant, the model structure would then need to be modified as shown in Figure 1.7. The length of river would have been divided into a finite difference grid and the changes in concentration between mesh points would be calculated using a numerical routine as shown in Chapter 3.

(f) *Production of computer program.* Since few steps are involved and few data, a very simple program will suffice:

Line No.	Instruction	Explanation
10	REMDATA	Label for data input
20	INPUT K	Input value for decay coefficient
30	READ A,B,C	Setting values of parameters
40	DATA 1,2,3	Numerical values for *A*, *B* and *C*
50	READ C2,Q1	Setting values of phenol conc.
60	DATA 0, 0.1	Corresponding numbers
70	FOR Q2 = 1 TO 2.5 STEP 0.5	Exploring the effect of various fresh-water flows
75	REM	Calculation of retention time
80	RT = A*B*(Q2)^C	Calculating the corresponding retention time

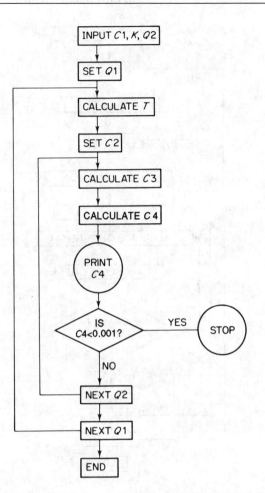

Figure 1.6. Flow diagram for simulation model

90	FOR C1 = 10 TO 50 STEP 5	Exploring the effect of various discharge concentrations
95	REM	Calculations of phenol concentration
100	C3 = ((Q1*C1) + (Q2*C2))/ (Q1 + Q2)	Calculates the concentration after mixing
110	C4 = C3*EXP(− K*RT)	Calculates concentration at abstraction point
115	REM OUTPUT	
120	PRINT "Q2 = ", Q2	Output
	PRINT "C1 = ", C1	
	PRINT "DOWNSTREAM PHENOL CONC = ", C4	

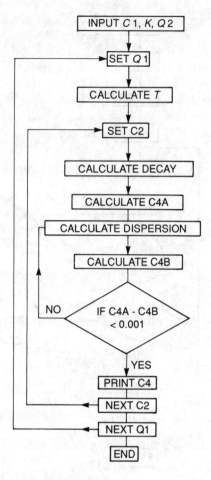

Figure 1.7. Modified model including dispersion

| 130 | NEXT C1 | End of loop |
| 140 | NEXT Q2 | End of loop |

(g) *Calibration of the model.* The parameters of the model A, B, C and K need to be evaluated. The first three are particular to the streams; these values may have been determined by government agencies but they may need to be experimentally determined by tracer studies in the stream. The evaluation of K can also be made experimentally by determining decay rates in phenol/river water mixtures over time. Alternatively, values of K can be taken from the literature if sensitivity analysis shows that this is not a crucial parameter.

The validation of the model depends upon local possibilities. Ideally one would like to test the output from the model against observations for several freshwater flows and discharge concentrations. This could be most easily simulated by gulp discharges of phenol at the site of the proposed discharge, or more elaborately by arranging a temporary pipeline.

(h) *Sensitivity analysis*. This should be used to determine the importance of the hydrological parameters (A, B and C) and the biochemical parameters.

(i) *The importance of temperature*. If the model were required to check this, it would require a small addition to the program to input a value for T and to correct K using an expression of the form $K_T = K\theta^{(T - 20)}$.

1.4 SIMULATION TERMINOLOGY

Many different types of water quality models are available so it is not possible to give a simple classification. However, the classification presented in Figure 1.8 may help to show some of the types of models that have been used, provided that it is appreciated that this is not an exhaustive list.

The fundamental distinction into simulation and optimization is in the model aims. Optimization is a group of mathematical techniques used to obtain the best (in some cases, for example the least costly) solution to some allocation situation. These models

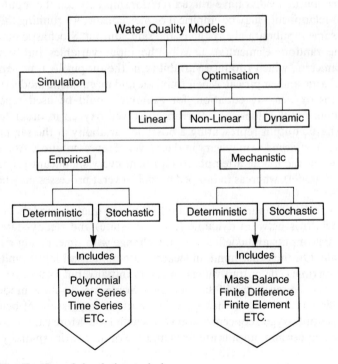

Figure 1.8. Classification of simulation techniques

have not been widely used in water pollution control and have not therefore been included in this book. There is however an example of a linear programming optimization in Chapter 7 and an example of a non-linear programming method in Chapter 11.

Simulation models can be mechanistic, empirical, deterministic, or stochastic. These are not neat exclusive classifications, for the elements of a complex model may contain examples of each. For example, in an estuary the dissolved oxygen concentration may be affected by variations in its freshwater flow. This might be represented by an empirical model, e.g.

$$Do = a + b_1 * \text{temperature} \pm b_2 * \log(\text{flow}) \pm b_3 * (\text{tidal amplitude})$$

(sometimes called a black-box model because it does not attempt to postulate any mechanism).

Various types of empirical relationships are widely used, three of which are listed in Figure 1.8, merely as illustration.

In mechanistic models there is an attempt to postulate some mechanism which relates the water quality parameter to the environmental. Figure 1.9 shows an example of a mechanistic model of dissolved oxygen in an estuary. This is the most frequently used type of water model and many variations have been devised. One notable subdivision is into the following classes:

(1) *Deterministic models* have a fixed relationship between the input and output. This relationship may be empirical or mechanistic. Re-running the model with the same input data always gives the same output. Stochastic models contain some random element(s), usually the input variables but sometimes the parameters, which produce variability in the output. A time series model is stochastic and empirical. A deterministic and mechanistic model (the Streeter–Phelps oxygen sag equation, for example) could be used repeatedly with random input values for upstream dissolved oxygen in order to generate a stochastic output representing a possible variability at the sag point.

(2) *Mechanistic models* may vary in their level of representation. At one end of the spectrum the model attempts to represent every significant process (a distributed model), whereas in lumped models several processes may be combined.

There is a further important distinction in water quality models not shown in Figure 1.8, which is between dynamic (unsteady-state) and steady-state models. In dynamic models inputs and coefficients may change with time, giving rise to a time-variant output. On the other hand, in steady-state models all inputs and coefficients are constant in time. The output values may vary initially, but as the system comes to equilibrium a fixed output will be obtained Some steady-state models, e.g. the Streeter–Phelps oxygen sag model, give the spurious impression of being dynamic because the output oxygen concentration varies with time. More careful consideration shows that all inputs and outputs are fixed and the output varies spatially rather than temporally.

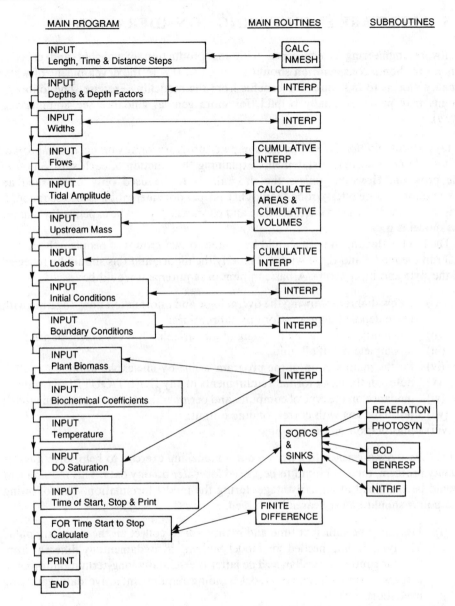

MAIN PROGRAM   MAIN ROUTINES   SUBROUTINES

INPUT
Length, Time & Distance Steps   CALC
                                       NMESH

INPUT
Depths & Depth Factors   INTERP

INPUT
Widths   INTERP

INPUT
Flows   CUMULATIVE
                                       INTERP

INPUT
Tidal Amplitude   CALCULATE
                                       AREAS &
                                       CUMULATIVE
INPUT   VOLUMES
Upstream Mass

INPUT   CUMULATIVE
Loads   INTERP

INPUT
Initial Conditions   INTERP

INPUT
Boundary Conditions   INTERP

INPUT
Plant Biomass   INTERP

INPUT
Biochemical Coefficients

INPUT   REAERATION
Temperature   PHOTOSYN
                                       SORCS
                                       &   BOD
INPUT   SINKS   BENRESP
DO Saturation
                                                     NITRIF
INPUT
Time of Start, Stop & Print   FINITE
                                       DIFFERENCE
FOR Time Start to Stop
Calculate

PRINT

END

Figure 1.9. Flow diagram of Hooghly estuary model

1.5 SOFTWARE ENGINEERING CONSIDERATION

Software engineering is concerned with the production of efficient and effective programs. Some consideration should be given to this in the development of water quality models to maximize the benefits from the modelling exercise. The following points may prove especially helpful (for more general guidance see Sommerville, 1989).

(a) *Documentation.* In writing the computer program for any model it is important to include non-executable statements explaining the function of each sub-section of the program. However, unless the program is to be used only once (such an assumption is more often wrong than right), proper documentation is also required. If this is neglected, then considerable time and effort will need to be expended whenever the model is used.

The level of documentation should be geared to two groups of people—those who will run the model and those who will modify the model, and this should be reflected in the style and its contents. A basic minimum requirement would be:

   (i)   a flow diagram showing the overal logic and information flow, together with more detailed diagrams for any subroutines;
  (ii) a glossary showing the meaning of all variables and coefficient names;
 (iii) a complete list of all units;
 (iv) for the main section of the program, a line-by-line explanation;
  (v) notes on the exact format requirements of any INPUT/OUTPUT routines;
 (vi) guidance on the type of computer and peripherals required to run the model;
 (vii) test data sets with corresponding output;
(viii) information on the run times.

(b) *Data requirements.* In carrying out a modelling exercise to help solve a water quality problem, there is bound to be a need for water quality data. This requirement should be examined at an early stage during the model formulation. The following two points should be very carefully studied.

  (i)   Will there be sufficient time and manpower to collect all the necessary data? The type of data needed for model building is fundamentally different from that for routine surveillance. The latter is essentially long-term monitoring in a sparse manner whereas model building requires intensive data gathering over short periods.
 (ii) Data (or lack of them) are likely to determine the success or failure of a modelling exercise. The manpower requirements for the formulations, calibration and validation of the model is likely to be less than five per cent of the manpower required to collect the data. In the case of hydraulically complex environments, model building may be less than one per cent of the total effort. The number and quality of available data and the resources available to obtain additional data need to be carefully estimated.

(c) *Errors*. Considerable attention should be payed in the design of the program to the avoidance of errors and the provision of advice when errors do occur. Many errors of the type that arise during compilation, reading files from disk, outputting to screen or plotter, etc., are due to incompatibilities between software and hardware. These can be avoided by detailed specification of the hardware requirements. Running a water quality program for the first time should always include a check with a test data set and the corresponding output. This limits any further errors to two categories: (1) typing errors or inconsistencies in the data set and (2) inappropriate values of coefficients or excessive size of data items. The search for these in large data files can be frustrating. Try to avoid these problems by providing the user with some input template that draws attention to inconsistencies or unusual values. Another issue which needs careful consideration is the choice of appropriate units. Inconsistency in the units is one of the commonest sources of error in model output.

REFERENCE

Sommerville, I. (1989). *Engineering Software*. Addison-Wesley, Wokingham.

APPENDIX A

The following program is a simple model of dissolved oxygen in a stream based on the 'oxygen sag' concept outlined in Chapters 5 and 7.

The variable names have the following meanings:

D1 = U/S DO   :B1 = U/S BOD   :Q1 = U/S flow
D2 = effluent DO   :B2 = effluent BOD   :Q2 = effluent flow
D3 = combined DO :B3 = combined BOD :Q3 = combined flow

A   = cross-sectional area
K1   = BOD decay coefficient
K2   = reaeration coefficient
D9   = DO saturation concentration
D(T)   = oxygen deficit at time T
D2(T)   = DO at time T
D4(T)   = distance downstream at time T

List of Dosag Model
```
10   REM SET UP ARRAYS
20   DIM D(50),D2(50),D4(50)
30   REM INPUT DATA (UNITS OF G,M,HR)
40   REM RIVER U/S OF DISCHARGE
50   Q1 = 5000
60   B1 = 2
```

```
 70   D1 = 8
 80   REM EFFLUENT DATA
 90   Q2 = 1000
100   B2 = 100
110   D2 = 2
120   REM OTHER DATA
130   A = 6
140   K1 = 0.1
150   K2 = 0.15
160   REM CALCULATE INITIAL CONDITIONS
170   D3 = (D1*Q1 + D2*Q2)/(Q1 + Q2)
180   B3 = (B1*Q1 + B2*Q2)/(Q1 + Q2)
190   REM CALCULATE DO DEFICIT Z
200   D9 = 10
210   Z = D9-D3
220   REM CALCULATE DEFICIT DOWNSTREAM
230   FOR T = 1 TO 20
240     D(T) = (K1*B3)/(K2-K1)
250     D(T) = D(T)*(EXP(-K1*T)-EXP(-K2*T))
260     D(T) = D(T) + Z*(EXP(-K2*T))
270     D2(T) = D9-D(T)
280     D4(T) = ((Q1 + Q2)*T)/A
290   NEXT T
300   REM OUTPUT DO PROFILE
310   PRINT" DISTANCE(M) DO(MG/L)"
320   FOR T = 1 TO 20
330     PRINT D2(T),TAB(8)D4(T)
350   NEXT T
```

```
> RUN
    DO(MG/L)            DISTANCE (M)
5.79979654        1000
4.92075247        2000
4.30347969        3000
3.89825673        4000
3.66354975        5000
3.56475189        6000
3.57310974        7000
3.66480978        8000
3.82020126        9000
4.02313588        10000
4.26040719        11000
 4.5212748        12000
4.79706134        13000
5.08081107        14000
```

5.36700136	15000
5.65129877	16000
5.93035354	17000
6.20162636	18000
6.46324276	19000
6.71387092	20000

Chapter 2

Introduction to Computing

A. JAMES AND S. MARKHAM

2.1 INTRODUCTION

All but the simplest of mathematical models requires the use of a computer, so a degree of computer literacy is required in water quality modelling. For someone who merely wishes to run existing models, this requires no more than a working knowledge of an operating system (e.g. MS-DOS, PC-DOS, OS2 or UNIX). But to understand how water quality models work it is necessary to learn some programming language. There are many languages, all with their adherents and advantages. For the purposes of this book the language that has been adopted is BASIC, not because it is fundamentally superior, but because it will be familiar to many readers and it can be explained fairly briefly to the others. BASIC is really a family language rather than a single language. The fundamental difference is between

(a) Compiled forms like QUICK BASIC where the whole program is converted (i.e. compiled) from the written form into machine code before it is executed. This helps with detecting errors in the program during compilation, permits the program to run much faster and allows the programmer to hand over compiled copies of a program which cannot be altered.

(b) Uncompiled forms, in which each statement of the program is interpreted each time it is encountered. These have the advantage of simplicity but can become slow to run when used for very large programs.

This chapter discusses those forms of BASIC, such as QUICK BASIC and BBC BASIC, that support structures such as procedures; there are other dialects that require the user to resort to subroutines involving GOSUB...RETURN commands.

An Introduction to Water Quality Modelling. 2nd Edition. Edited by A. James

Later in this chapter there are some notes that provide a brief introduction to uncompiled BASIC. This is one of the most portable of languages (i.e. runs on a variety of machines) but it must be appreciated that there is a bewildering variety of computers available and not all forms of BASIC will work on all machines. Up until the early 1980s the costs of hardware (computers, printers, etc.) dictated the choice of software. This situation has changed and the choice of computer is often dictated by its compatibility with popular software packages. Advice on the choice of hardware is so ephemeral as to be useless, but along with other considerations like size, cost, servicing do not neglect software compatibility in choosing a computer for water quality modelling.

Apart from the operating system (DOS) and the programming language (BASIC) there is one other programming technique which is increasingly being used in water quality modelling. This is spreadsheets. As with programming languages there is a wide variety of spreadsheets available, ranging from very simple packages like AS-EASY-AS to sophisticated ones like LOTUS. These provide the most attractive of the three options for creating input and output files for water quality modelling. The other alternatives are as follows:

(a) *Word processors.* There are a tremendous number of word-processing packages available, many of which (e.g. WORDSTAR) provide an editor. This editor can be used to create and amend data files for input/output into a model.

(b) *Editors.* Most mainframe computers and most personal computers have editorial packages (e.g. IBM PROFESSIONAL EDITOR) which can be used to create and edit files.

The advantage of the more sophisticated spreadsheet packages is that the output from the model can be manipulated for example into graphical format. Alternatively graphical routines can be built into the model.

2.2 OPERATING SYSTEM

The operation of a computer is controlled by an operating system which may be permanently installed on a hard disk or temporarily installed by inserting a floppy disk. Operating systems are complex and sophisticated enabling the user to carry out a wide variety of tasks but, for the purposes of developing and running water quality models, this may be reduced to an understanding of the following few commands:

COMMAND		MEANING
(i)	A: OR C:	Changes the disk controlling operation of the computer to that installed in the specified drive.
(ii)	DIR	Displays the list of files on the disk in the current directory.
(iii)	CD\SAM	Changes directory to the directory named after the back slash—in this case sub-directory SAM.
(iv)	MKDIR SAM	Creates a new directory with the name specified.

(v)	FORMAT	Prepares a disk so that it can be used for reading from and writing to. This is necessary for all new disks and when re-using a disk. But be careful it removes all existing programs.
(vi)	COPY.	Copies programs from one disk to another or one directory to another.
(vii)	DEL	Deletes a file from a disk. This is used to make space for other files.
(viii)	TYPE	Types the contents of the file on the screen or can be redirected to the printer.

The commands given here are specific to DOS. For greater detail and additional commands the reader should consult the DOS manual.

Note that some computers have a built-in operating system which function by choosing icons (symbols) from menus on the screen using a mouse rather than a keyboard. The mouse is used to move the screen pointer to the desired menu or task and clicked to indicate the choice.

2.3 PROGRAMMING IN BASIC

Programs consist of a series of instructions together with the data necessary to carry out these instructions (INPUT). The results of the calculation (OUTPUT) are then displayed.

Most personal computers have at least two disk drives and one of these usually operates a hard disk that is built into the machine. The hard disk is used to store the operating system, the editor, any compilers and frequently used programs and data files. These are arranged on the hard disk in a hierarchical manner. The aim of the positioning is to classify files into useful groupings. For example, if you regularly use both FORTRAN and BASIC and are working on water quality models for lakes and rivers, then you could use the following arrangement:

MAIN DIRECTORY	Containing operating systems, word processor etc.
SUBDIRECTORY FORTRAN	Containing Fortran compiler, editor etc.
	SUB-SUBDIRECTORY, River
	SUB-SUBDIRECTORY, Lakes
SUBDIRECTORY	BASIC containing BASIC interpreter etc.
	SUB-SUBDIRECTORY, River
	*SUB-SUBDIRECTORY, Lakes

Different disk drives are referred to by letters often A, B, C, D, E. Generally only the hard disk is large enough to be worth partitioning.

Remember to back-up your important program and data files by keeping two copies: one on the hard disk and one on the floppy disk, stored for subsequent use. The INPUT data may be contained in the program but is more often stored in a separate data file.

A BASIC program consists of a series of lines, which will be executed in the order defined by the line numbers. You must therefore number each line but gaps in the line numbers are permissible, to allow line insertion. It is therefore customary to number lines in multiples of 10.

2.3.1 Variables and Input Data

The program should begin by supplying the information needed to carry out any instruction. For numerical programs this means entering the values of the variables being used. This can be done in various ways, e.g.

  (i) Assign values
      10 A% = 10
      20 BP8 = 4.01

  (ii) Use INPUT command
      10 INPUT A,C,DR5

this allows the values of variables to be assigned when the program is running. Each time you run the program a question mark will appear on the screen for each INPUT and you must type the numerical values of the variables (separated by commas) and press enter.

  (iii) Use READ command
      10 READ V,Z9%,CAF4
      20 DATA 5.2,24,101.2

For each READ there must be a corresponding DATA line which contains the values of the variables. Note that names of variables may be any combination of up to 31 alphanumeric characters beginning with a letter. When individual variables are used they do not need to be declared at the beginning of the program—a new variable can be introduced at any point provided that it is given a numerical value.

Integer variables are distinguished from floating point variables by the use of the % symbol, e.g.

    $A\%$ is an integer variable
    A is a floating point variable.

When you want to store a name instead of a number, this is called a string variable and you use the $ symbol, e.g.

    $B\$ = $ "JONES"

assigns the name Jones to the variable $B\$$.

For arrays some dimension statements are needed at the start of the program. This is usually done by

    10 DIM A (10), BOP (20), CGRA (5,4), B$ (6)

This dimensions a linear array of 10 for A and 20 for BOP, a two-dimensional array of 5×4 for $CGRA$ and a linear array of 6 for the string variable $B\$$.

When variables are used in programs that contain separate sections (called PROCEDURES) they need to be declared as global variables so that the computer will transfer their values to all parts of the program. The statement used is

    10 GLOBAL A%, B5

and this must appear in each PROCEDURE where the global variable is used. (Alternatively values can be passed into and out of procedure using a defined list of variables called arguments.)

Some dialects of BASIC do not support PROCEDURES as outlined here. These employ a simpler structure known as a subroutine, which is simply a small fragment of program called by the main program, control being returned to the calling point after execution.

2.3.2 Calculations and Iterations

Once the variables have been defined the programmer can then specify the calculations to be carried out. These are written using the following symbols.

+ Add	* Multiply	^Raise to power
− Subtract	/ Divide	\ Integer division
		(sometimes MOD is used rather than \\)

plus a large number of special functions such as

SIN sine	EXP exponential e^x
LN logarithm	INT integer

NOTE: LN gives natural logarithm
           Log to the base 10 can be obtained by using LOG

The calculation should be written as a series of instructions. Complex calculations can be broken down and written on a number of lines, e.g.

$$A = \frac{B(C^2 + D)}{Z\left(\dfrac{H^3}{2} + \sqrt{PR}\right)}$$

can be written as

    10 NUM = (C^2 + D)*B
    20 DENOM = Z*((H^3)/2 + SQR(P*R))
    30 A = NUM/DENOM

The computer executes lines in the order specified by the line numbers. Within each line the priority rules are as follows:

(i)   Function, exponentiation, multiplication or division, addition or subtraction.

(ii) Where priorities are equal it works from left to right but these can be overridden by the use of brackets, e.g.

10 A = 1	10 A = 1
20 B = 10	20 B = 10
30 C = 20 + A*B	30 C = (20 + A)*B
C = 30	C = 210

Where brackets are nested the contents of the innermost brackets are calculated first and the computer then works outwards.

In some situations the program can be written to carry out a complex calculation for one set of data. The program can then be re-run using a modified data set, especially when data are being entered from the keyboard using an INPUT command. But it is usually more convenient to make the program run through the calculations several times by using an iteration command. There are various commands available; those most often used are

FOR $\cdots$ NEXT
REPEAT $\cdots$ UNTIL

If it is necessary to produce a sine wave to simulate some natural phenomenon this could be produced as the value of X for each hour through a diurnal cycle using the following procedure:

```
10  FOR T = 1 TO 24
20      X = SIN(T/4)
30      PRINT X
40  NEXT T
```

In the above example the variable T is used in the calculation and is incremented at each iteration. More often the variable used in the FOR statement is merely a counter, e.g.

```
10  FOR N = 1 TO 20
20      READ A
30      B = A/4*A^2
40      PRINT B
50  NEXT N
```

The alternative REPEAT $\cdots$ UNTIL is still used in a slightly different context where some variable is being changed until a certain result is obtained, e.g.

```
10  REPEAT
20      READ A
30      B = A/4 + A^2
40      PRINT B
50  UNTIL B > 1000
```

The FOR···NEXT combination is the most versatile since the step size can be varied, e.g.

FOR T = 1 TO 24 STEP 2   or

FOR T = 24 TO 1 STEP-1

and it can achieve the same effect as REPEAT··· UNTIL when an IF command is introduced.

Remember to leave a space after all special words like FOR, TO, NEXT, etc.

In the above example the program would become

```
10  FOR N = 1 TO 20
20      READ A
30      B = A/4*A^2
40      PRINT B
50      IF B > 1000 THEN 150
60  NEXT N
    .
    .
    .
150 PRINT "B EXCEEDS 1000"
160 STOP
```

Note that this causes the program to jump out of the loop and the program execution terminates. Such procedures should be used with care and it is inadvisable to use similar techniques such as GOTO to jump into loops as the values of variables may not be properly assigned.

One useful command used in connection with iteration is RESTORE. This enables a data set to be re-used several times, e.g.

```
30  FOR N = 1 TO 5
40      READ A,B
50      C = (N + A)*B
60      PRINT C
70      RESTORE
80  NEXT N
```

2.3.3   Subroutines

Problems for programming are often more manageable when broken down into smaller parts especially when some of the calculations are frequently repeated.

An example is in the design of a pipe network for storm drainage. The maximum' capacity of a pipe is defined by the equation

$$Flow = A*[- 2*SQR(2*G*D*S)*LOG(CK/(3.7 + D)) + (2.2*V))/$$

$$(D*SQR(2*D*G*S))]$$

In the design of a network this calculation will be used many times. It is therefore good technique to write the solution of this equation as a separate subroutine and simply call this whenever it is needed from the main programme.

In BASIC such a subroutine is referred to as a PROCEDURE. In the program it is simply called by giving the name. Somewhere else in the program it is defined by:

> DEFPROC
> Contents
> ENDPROC

This returns the computer to the program at the next line after the subroutine was called.

Complex programs consist of a main program with a large number of such PROCEDURES. This helps in debugging programs since each PROCEDURE can be tested separately and also helps to show the flow of information.

The flow diagram is shown in Figure 2.1.

The values of variables used in PROCEDURES need to be defined. This can be done by passing the values through the arguments,

e.g. 102 PIPESIZE (G,V,R)

These variables are then referred to as GLOBAL, i.e. they are common to both the main program and the subroutine. The only restriction is that arrays cannot be passed in this way. Arrays are assumed to be global and cannot be treated as local.

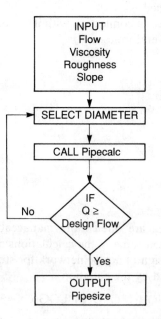

Figure 2.1. Flow diagram for pipe calculation

2.3.4 Output

Having completed the calculations it is necessary to obtain the values that have been calculated either on the screen or on the printer.

Using the command PRINT will display the current numerical or string value of a variable. Note that when a programme is iterating the value of a variable is changing each time, so that it is necessary to put the PRINT command inside the iteration loop.

When several variables are being displayed it is often useful to label them to avoid confusion using inverted commas, e.g.

    50 PRINT "X = ", X

will display X = (current numerical value).

When tables of values are to be printed the table can be labelled early on in the program using

    20 PRINT "X   Y   Z"

and then the variable values printed out inside a FOR ⋯ NEXT loop

    80 FOR N = 1 TO 10
    .

    .

    .
    190   PRINT X,Y,Z
    200 NEXT N

To obtain output from the printer use the VDU2 (or other specialist command) in front of the output required.

To list the program

    50 VDU 2
    60 LIST

To print results

    30 VDU 2
    40 PRINT B,C,D

To change back to the screen from the printer use

    50 VDU3
    60 PRINT B,C,D

2.3.5 Running and Editing a Program

Once a program has been written it can be tested and used with the RUN command. On a good day the program will be executed successfully and no error messages will be obtained, but more often error will cause the execution to be aborted or the program will run with some numerical errors that cause the output to be incorrect. In the latter case it is possible to stop the execution using Control Break. In either case it is necessary to edit the program to correct the errors by first displaying the program using the LIST command.

LIST—lists the whole program, but if this is too long to fit on the screen use LIST TO 100 (which lists all lines up to 100) or LIST 100 TO (which lists all lines from 100 to the end of the program). When the lines containing the errors are detected these can be amended using the EDIT command.

After displaying the appropriate line or lines on the screen use the cursor controls to take the cursor to the beginning of the line to be edited. Press the COPY key to copy the line and use the other keys to replace or insert characters or the cursor key to miss out characters. When the line has been amended press the ENTER key.

If you are editing a line and decide you wish to retain the original version, press the ESC key.

2.3.6   Saving and Loading a Program

Once a program has been created it can be saved on a disk for future use with the SAVE command. The program receives its name at the same time, e.g.

SAVE "BILL"

and then stores the current program on disk under the name BILL. It may be recovered at any time by inserting the disk and typing LOAD "BILL". You can check which files are stored on a disk by using the DIR(ectory) command.

Remember that programs may be stored only on formatted disks and that formatting a disk erases all previous programs.

2.3.7   Finishing

To leave BASIC and return to the normal operating system type SYSTEM or in some versions of BASIC type *BYE.

2.4   INTRODUCTION TO SPREADSHEETS

These few paragraphs give a taste of spreadsheets. The construction of spreadsheet programs requires a reference manual. There are many commercial spreadsheet packages available but they are all very similar in composition and use. Spreadsheets have come into their own through their ease of use and the ability to assemble a program in stages, verify and amend from the same environment.

A spreadsheet is fundamentally an electronic sheet of paper that is divided into cells. Each cell is referenced by a row (designated by a number) and a column (designated by a letter); the cell is usually capable of storing up to 260 characters, though these do not all have to appear on the screen at once.

A cell can be one of two types—a 'label' or 'value'. A label cell allows the annotation of the model with informative text. A value cell can be a number or a formula.

For example consider the velocity in a pipe (V):

$$V = \frac{Q}{A}$$

Q = discharge in pipe
A = cross-sectional area of pipe.

This can be represented in a spreadsheet as follows:

	A		B	C	D	E	
1							
2	Discharge (Q)	=		1.5 m3/s		number	
3							
4	Radius of pipe	=		0.8 m		number	
5							
6	Area of pipe (A)	=		2.011 m2		@PI*(C4)^2	
7							
8	Pipe velocity (V)	=		0.746 m/s		+ C2/C6	
9							

The calculations for the pipe discharge is done in *Column C*. A number can be entered from the keyboard into *CELLs C2* or *C4* and the pipe area and velocity is immediately calculated in *CELLS C6* and *C8* respectively. The *CELLs* in *column E* show the contents of the *CELLs* in *column C*, for example the formula in *CELL C6* is shown in *E6*. The cells in *column A* are labels; they overrun into the *B column* because those cells contain nothing.

Spreadsheets have powerful edit facilities which allow the movement and copying of cells. Graphic and database capability is usually included, and a large library of functions from logarithms to net present value (NPV). (The function used in the example above was @PI for the value of π.)

Within a spreadsheet is a macro programming language that allows the process of printing to a printer to be performed with one keystroke. This macro language enables the construction of programs that would normally have been written in a language such as BASIC.

Chapter 3

Introduction to Numerical Methods

D. J. ELLIOTT

3.1 INTRODUCTION

The process of formulation of a water quality model may be broken down into a discrete number of steps, the first two of which are:

(1) Identify the physical, chemical and biochemical laws which govern the system under consideration.
(2) Express these laws in a precise mathematical form.

Frequently, in water quality models, the equations produced in step (2) are not amenable to direct analytical solution and simplifying assumptions are made to reduce the complexity of the problem. Often the equations are reduced to a level at which an analytical solution is possible. However, even with a simplified formulation it may be more convenient to use a numerical rather than analytical approach to finding a solution.

This chapter introduces some of the numerical techniques which may, with the aid of a computer, form the basis of water quality models.

3.2 SYSTEMS OF SIMULTANEOUS LINEAR EQUATIONS

Linear equations are used to model many different types of system. The set of equations may be the direct result of the way in which the model is formulated such as in the large economic models with many variables, or the equations may be developed indirectly through use of numerical analysis to solve problems. For example, the

---

An Introduction to Water Quality Modelling. 2nd Edition. Edited by A. James

solution of differential equations by finite difference methods and statistical regression analysis both produce systems of linear equations which require solution.

3.2.1 Linearity and the Existence of Solutions

A linear equation contains in each term only one variable and each variable appears only to the first power.

$$x + y - 3z = 1 \qquad \text{is linear} \tag{3.1}$$

$$xy + 3z - 9 = 9 \qquad \text{is not linear} \tag{3.2}$$

$$x^2 - y - z = 11 \qquad \text{is not linear} \tag{3.3}$$

A solution exists to a system of linear equations if on substituting a set of values for all variables in the systems, all equations are satisfied simultaneously.

$$x + y + z = 9 \tag{3.4}$$

$$2x - 3y + z = 1 \tag{3.5}$$

$$x - y + 3z = 7 \tag{3.6}$$

This set of equations has the solution $x = 4$, $y = 3$, $z = 2$.

It is usually not possible to say without detailed examination of the equations if there is a solution and if it is unique.

Three possibilities exist. Consider Figure 3.1

$$\text{Line A is represented by } 3y - 2x + 1 = 0$$

$$\text{Line B is represented by } y + 2x - 5 = 0$$

$$\text{Line C is represented by } y + 2x - 12 = 0$$

(1) The system

$$\left.\begin{array}{l} 3y - 2x + 1 = 0 \\ y + 2x - 5 = 0 \end{array}\right\} \tag{3.7}$$

has the solution $x = 2$, $y = 1$ which is unique.

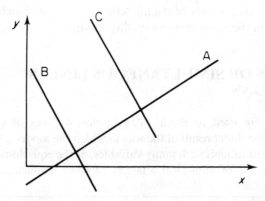

Figure 3.1. Graph of three simultaneous equations

(2) The system

$$y + 2x - 5 = 0 \atop y + 2x - 12 = 0 \biggr\}$$

(3.8)

has no solution. The lines are parallel and because they never meet they have no solution.

(3) The system

$$y + 2x - 5 = 0 \atop 2y + 4x - 10 = 0 \biggr\}$$

(3.9)

has an infinite number of solutions because the equations describe the same line and any point on the line is a solution.

Systems (2) and (3) are said to be singular. It is necessary to test for singularity which may be done directly using the determinant of the system or indirectly through the particular method of solution adopted.

Singularity of a system of equations is indicated if the determinant of the system coefficients is zero.

3.2.2 Determinants

The solution to the following system of equations

$$a_1 x + b_1 y = c_1 \quad \text{(a)} \atop a_2 x + b_2 y = c_2 \quad \text{(b)} \biggr\}$$

(3.10)

may be found by finding y in terms of x in equation (3.10b) and substituting in equation (3.10a)

$$a_1 x + \frac{b_1}{b_2}(c_2 - a_2 x) = c_1$$

(3.11)

rearranging gives

$$x = \frac{c_1 b_2 - c_2 b_1}{a_1 b_2 - a_2 b_1}$$

(3.12)

Providing $a_1 b_2 - a_2 b_1 \neq 0$ a unique solution for x is available.

$a_1 b_2 - a_2 b_1$ may be rewritten as $\begin{vmatrix} a_1 & b_1 \\ a_2 & b_2 \end{vmatrix}$ and

is called a second-order determinant (two rows and two columns).

Thus, using this notation

$$x = \frac{\begin{vmatrix} c_1 & b_1 \\ c_2 & b_2 \end{vmatrix}}{\begin{vmatrix} a_1 & b_1 \\ a_2 & b_2 \end{vmatrix}} \quad y = \frac{\begin{vmatrix} c_1 & a_1 \\ c_2 & a_2 \end{vmatrix}}{\begin{vmatrix} a_1 & b_1 \\ a_2 & b_2 \end{vmatrix}}$$

(3.13)

or

$$\frac{x}{\begin{vmatrix} c_1 & b_1 \\ c_2 & b_2 \end{vmatrix}} = \frac{y}{\begin{vmatrix} c_1 & a_1 \\ c_2 & a_2 \end{vmatrix}} = \frac{1}{\begin{vmatrix} a_1 & b_1 \\ a_2 & b_2 \end{vmatrix}} \tag{3.14}$$

A system of three equations in three unknowns

$$a_1 x + b_1 y + c_1 z = d_1$$

$$a_2 x + b_2 y + c_2 z = d_2 \tag{3.15}$$

$$a_3 x + b_3 y + c_3 z = d_3$$

has the solution

$$\frac{x}{\begin{vmatrix} b_1 & c_1 & d_1 \\ b_2 & c_2 & d_2 \\ b_3 & c_3 & d_3 \end{vmatrix}} = \frac{y}{\begin{vmatrix} a_1 & c_1 & d_1 \\ a_2 & c_2 & d_2 \\ a_3 & c_3 & d_3 \end{vmatrix}} = \frac{z}{\begin{vmatrix} a_1 & b_1 & d_1 \\ a_2 & b_2 & d_2 \\ a_3 & b_3 & d_3 \end{vmatrix}} = \frac{1}{\begin{vmatrix} a_1 & b_1 & c_1 \\ a_2 & b_2 & c_2 \\ a_3 & b_3 & c_3 \end{vmatrix}} \tag{3.16}$$

where

$$\begin{vmatrix} a_1 & b_1 & c_1 \\ a_2 & b_2 & c_2 \\ a_3 & b_3 & c_3 \end{vmatrix}$$

is defined to mean

$$a_1 \begin{vmatrix} b_2 & c_2 \\ b_3 & c_3 \end{vmatrix} - a_2 \begin{vmatrix} b_1 & c_1 \\ b_3 & c_3 \end{vmatrix} + a_3 \begin{vmatrix} b_1 & c_1 \\ b_2 & c_2 \end{vmatrix}$$

or

$$a_1(b_2 c_3 - c_2 b_3) - a_2(b_1 c_3 - c_1 b_3) + a_3(b_1 c_2 - c_1 b_2)$$

The other determinants may be evaluated in a similar manner.

Note that in equation (3.16) the determinant of the last fraction is obtained from writing the coefficients in the form in which they appear on the left-hand side of the original equations (3.15). This is a determinant of the third order.

The leading element of the determinant is a_1. If the row and column containing the element is omitted then the determinant of the remaining elements is called the minor of that element.

e.g., the minor of a_3 is $\begin{vmatrix} b_1 & c_1 \\ b_2 & c_2 \end{vmatrix}$

As a general rule, higher orders of determinants can be evaluated by multiplying the elements in the first column by their minors. Starting with the leading element attach $+$ and $-$ signs alternately to these products.

3.2.3 Cofactors

The $+$ and $-$ signs associated with the minors may be set out as shown in Figure 3.2. If the signs in Figure 3.2 are attached to the corresponding minors the cofactors of the

$$\begin{vmatrix} + & - & + & - & \cdots \\ - & + & - & + & \cdots \\ + & - & + & - & \cdots \\ - & + & - & + & \cdots \\ \vdots & \vdots & \vdots & \vdots & \end{vmatrix}$$

Figure 3.2.

elements are obtained denoted by the capital letter corresponding to the element.

$$\begin{vmatrix} a_1 & b_1 & c_1 \\ a_2 & b_2 & c_2 \\ a_3 & b_3 & c_3 \end{vmatrix}$$

$$A_2 = - \begin{vmatrix} b_1 & c_1 \\ b_3 & c_3 \end{vmatrix} \quad B_2 = + \begin{vmatrix} a_1 & c_1 \\ a_3 & c_3 \end{vmatrix} \quad C_2 = - \begin{vmatrix} a_1 & b_1 \\ a_3 & b_3 \end{vmatrix}$$

Determinants may be evaluated by expanding along any row or column by multiplying each element in the row or by its cofactor and adding the products.

Referring again to equation (3.16) it can be seen that the solution for x given by

$$x = \frac{\begin{vmatrix} b_1 & c_1 & d_1 \\ b_2 & c_2 & d_2 \\ b_3 & c_3 & d_3 \end{vmatrix}}{\begin{vmatrix} a_1 & b_1 & c_1 \\ a_2 & b_2 & c_2 \\ a_3 & b_3 & c_3 \end{vmatrix}} \tag{3.17}$$

is based purely on the coefficient of the original system equation (3.15). It is therefore possible, when using computers for the solution to simultaneous equations, to store and manipulate the coefficients separately from the variables. It is also useful to use matrix notation to write the equations in shorthand form.

The general system of n variables and n unknowns may be written as

$$\left. \begin{array}{l} a_{11}x_1 + a_{12}x_2 + a_{13}x_3 + \cdots + a_{1n}x_n = b_1 \\ \qquad\qquad\qquad \vdots \\ a_{n1}x_1 + a_{1n}x_2 + a_{13}x_3 + \cdots + a_{nn}x_n = b_n \end{array} \right\} \tag{3.18}$$

The coefficients of the variables may be written as a matrix

$$A = \begin{bmatrix} a_{11} & a_{12} & \cdots & a_{13} \\ \vdots & \vdots & a_{ij} & \vdots \\ a_{n1} & a_{n2} & \ddots & a_{nn} \end{bmatrix} \leftarrow \text{row } i$$

$$\text{Column } j$$

and the coefficient a_{ij} denotes the coefficient of x_j in the ith equation.

The constants b_i may be written as a column vector

$$b = \begin{bmatrix} b_1 \\ b_2 \\ b_3 \\ \vdots \\ b_n \end{bmatrix}$$

hence the system equation (3.18) may be written as

$$Ax = b \tag{3.19}$$

Where A is a matrix and x and b are column vectors. Provided A is non-singular, equation (3.19) may be solved by writing

$$x = A^{-1}b \tag{3.20}$$

where A^{-1} is the inverse of A and from matrix algebra

$$A^{-1}A = I$$

where I is the identity matrix with ones on the principal diagonal and zeros elsewhere

$$I = \begin{bmatrix} 1 & 0 & 0 & \cdots & 0 \\ 0 & 1 & 0 & \cdots & 0 \\ 0 & 0 & 1 & \cdots & 0 \\ \vdots & \vdots & \vdots & & \vdots \\ 0 & 0 & 0 & \cdots & 1 \end{bmatrix} \tag{3.21}$$

A^{-1} may be defined as

$$A^{-1} = \left[\frac{C}{|A|} \right] \tag{3.22}$$

where

(1) C is the adjoint of A and is the transpose of the matrix of cofactors (i.e., the rows and columns have been interchanged).

$$C = \begin{bmatrix} A_{11} & A_{21} & A_{31} \\ A_{12} & A_{22} & A_{32} \\ A_{13} & A_{23} & A_{33} \end{bmatrix}$$

(2) $|A|$ is the determinant of A.
(3) A_{11}–A_{33} are the cofactors of A.

Hence

$$x = \left[\frac{C}{|A|} \right] \times b \tag{3.23}$$

Taking a numerical example by writing (4), (5) and (6) in the form of equation (3.19),

$$x_1 + x_2 + x_3 = 9$$
$$2x_1 - 3x_3 + x_3 = 1 \qquad (3.24)$$
$$x_1 - x_2 + 3x_3 = 7$$

$$A = \begin{bmatrix} 1 & 1 & 1 \\ 2 & -3 & 1 \\ 1 & -1 & 3 \end{bmatrix} \qquad b = \begin{bmatrix} 9 \\ 1 \\ 7 \end{bmatrix} \qquad C = \begin{bmatrix} A_{11} & A_{21} & A_{31} \\ A_{12} & A_{22} & A_{32} \\ A_{13} & A_{23} & A_{33} \end{bmatrix}$$

$$|A| = 1 \times (-3 \times 3 - 1 \times -1) - 1 \times (2 \times 3 - 1 \times 1)$$
$$+ 1 \times (2 \times -1 - -3 \times 1) = -12$$

$$A_{11} = + \begin{vmatrix} -3 & 1 \\ -1 & 3 \end{vmatrix} = -8$$

$$A_{12} = - \begin{vmatrix} 2 & 1 \\ 1 & 3 \end{vmatrix} = -5$$

$$A_{13} = + \begin{vmatrix} 2 & -3 \\ 1 & -1 \end{vmatrix} = 1$$

The other cofactors may be similarly evaluated to give

$$C = \begin{bmatrix} -8 & -4 & 4 \\ -5 & 2 & 1 \\ 1 & 2 & -5 \end{bmatrix}$$

Thus

$$x = \frac{\begin{bmatrix} -8 & -4 & 4 \\ -5 & 2 & 1 \\ 1 & 2 & -5 \end{bmatrix}}{-12} \begin{bmatrix} 9 \\ 1 \\ 7 \end{bmatrix} \qquad (3.25)$$

x_1, x_2 and x_3 may be evaluated using the rule for matrix multiplication, i.e., the element in each row of the first matrix is multiplied by the corresponding element in the column of the second matrix

$$x_1 = (-8 \times 9 + -4 \times 1 + 4 \times 7)/ -12 = +4$$
$$x_2 = (-5 \times 9 + 2 \times 1 + 1 \times 7)/ -12 = +3 \qquad (3.26)$$
$$x_3 = (1 \times 9 + 2 \times 1 - 5 \times 7) - 12 = +2$$

This method of solving equations is known as Cramer's rule. It is not generally considered useful as a numerical technique because it involves the calculation of n^2 determinants, each having order $n - 1$.

There are many methods available for solving systems of linear equations; most algorithms are designed in some way to minimize computer time or allocate storage efficiently for large systems in which many of the coefficients have zero value.

Two methods will be described here to illustrate the concepts on which more sophisticated packages are based.

3.2.4 Gauss Elimination Method

Using matrix algebra it is possible to transform the system of equations into an equivalent system of triangular form which can be solved by a back substitution process.

$$x_1 + x_2 + x_3 = 9 \quad \text{(A)}$$
$$2x_1 - 3x_2 + x_3 = 1 \quad \text{(B)} \tag{3.27}$$
$$x_1 - x_2 + 3x_3 = 7 \quad \text{(C)}$$

The first step is to eliminate x_1 from equations (3.27) (B) and (C)

$$x_1 + x_2 + x_3 = 9 \qquad \text{(A}_1\text{)}$$
$$0 - 2.5x_2 - 0.5x_3 = -8.5 \quad \text{(B}_1\text{)} \quad \text{((B)/2} - \text{(A))}$$
$$0 - 2x_2 + 2x_3 = -2 \quad \text{(C}_1\text{)} \quad \text{((C)} - \text{(A))}$$

The second step is to eliminate x_2 from (C$_1$)

$$x_1 + x_2 + x_3 = 9 \quad \text{(A}_2\text{)}$$
$$0 + 2x_2 + x_3 = 8 \quad \text{(B}_2\text{)} \quad \text{((C}_1/4 - \text{(B}_1\text{))}$$
$$0 + 0 + 12x_3 = 24 \quad \text{(C}_2\text{)} \quad \text{((5x(C}_1\text{)} - 4x\text{(B}_1\text{))}$$

The coefficient matrix is now

$$\begin{bmatrix} 1 & 1 & 1 \\ 0 & 2 & 1 \\ 0 & 0 & 12 \end{bmatrix}$$

Which is an upper triangular matrix with all coefficients zero below the principal diagonal.

From (C$_2$) $\quad x_3 = 2$

From (B$_2$) $\quad 2x_2 + 2 = 8 \qquad x_2 = \dfrac{6}{2} = 3$

From (A$_2$) $\quad x_1 + 3 + 2 = 9 \qquad x_1 = 4$

In general system equation (3.27) may be expressed as

$$a_{11}x_1 + a_{12}x_2 + a_{13}x_3 = b_1 \quad \text{(a)}$$
$$a_{21}x_1 + a_{22}x_2 + a_{23}x_3 = b_2 \quad \text{(b)} \tag{3.28}$$
$$a_{31}x_1 + a_{32}x_2 + a_{33}x_3 = b_3 \quad \text{(c)}$$

with which may be associated the augmented matrix form as

$$
\begin{bmatrix}
a_{11} & a_{12} & a_{13} & b_1 \\
a_{21} & a_{22} & a_{23} & b_2 \\
a_{31} & a_{32} & a_{33} & b_3
\end{bmatrix}
\begin{array}{l}
(a) \\
(b) \\
(c)
\end{array}
\qquad (3.29)
$$

a_{21} may be eliminated by

(1) Specifying a multiplier $- a_{21}/a_{11}$
(2) Adding $- (a_{21}/a_{11}) \times$ row (a) (equation (3.29)) to row (b) (equation (3.29))

a_{31} may be eliminated by

(1) Specifying a multiplier $- a_{31}/a_{11}$
(2) Adding $- (a_{31}/a_{11}) \times$ row (a) (equation (3.29)) to (b) (equation (3.29)).

Hence a new matrix is formed

$$
\begin{bmatrix}
a_{11} & a_{12} & a_{13} & b_1 \\
0 & a'_{22} & a'_{23} & b'_2 \\
0 & a'_{32} & a'_{33} & b'_3
\end{bmatrix}
\begin{array}{l}
(a1) \\
(b1) \\
(c1)
\end{array}
\qquad (3.30)
$$

The values of elements in row (a1) (equation (3.30)) remain the same, elements in rows (b1) (equation (3.30)) and (c1) (equation (3.30)) such as a_{22} have a new value obtained from $a_{22} - a_{12}(a_{21}/a_{11})$ etc.
   The element a'_{32} may be eliminated by

(1) Specifying a multiplier $- a'_{32}/a'_{22}$

(2) Adding $- (a'_{32}/a'_{22}) \times$ row (b1) (equation (3.30)) to row (c1) (equation (3.30)).

The final system of equations is now

$$
a_{11}x_1 + a_{12}x_2 + a_{13}x_3 = b_1
$$
$$
0 + a'_{22}x_2 + a'_{23}x_3 = b_2 \qquad (3.31)
$$
$$
0 + 0 + a''_{33}x_3 = b_3
$$

which may be solved by back substitution

$$
x_3 = b_3/a''_{33}
$$
$$
x_2 = (b_2 - a'_{23}x_3)/a'_{22} \qquad (3.32)
$$
$$
x_1 = (b_1 - a_{13}x_3 - a_{12}x_2)/a_{11}
$$

The diagonal elements a_{11}, a_{22} and a_{33} are known as pivot elements and occur in the denominator of the multipliers and also in the back substitution process. To carry out the successive elimination procedure it is necessary for the pivot elements to be non-zero. If at any stage the pivot element is zero the remaining rows may be interchanged to produce a non-zero pivot. If it is not possible to find a non-zero pivot, the system of linear equations has no solution.

It is not possible here to discuss in great detail the accuracy of a numerical solution to a system of linear equations. However, some of the sources of error are indicated below:

(1)   The value of the pivot elements can affect the answer significantly. If the pivot element is small compared to other elements in its column which have to be eliminated then the multiplier will be greater than one in magnitude which may lead to an increase in round-off errors. It is possible to minimize this problem by arranging, at each elimination stage, the rows of the matrix below the pivot so that the new pivot is larger in absolute value than any element beneath it in its column. Thus multipliers have a magnitude less than or equal to one.

(2)   Uncertainty in the coefficients of the variables may also lead to errors particularly if the coefficients are obtained from experimental observation. This problem may even be more severe if the equations are ill-conditioned, in which case the solutions are very sensitive to small changes in coefficient value.

(3)   Round-off errors may be significant, particularly in large systems of equations. These errors are propagated at each step in the solution procedure. The growth of such errors can lead to completely useless results.

3.2.5   An Iterative Method

Large systems of linear equations cannot be solved with any degree of certainty by the direct approach described above. Several iterative methods have been developed, one of which is the Gauss–Seidel approach. As with all iterative methods, it starts with an initial approximate solution and uses this in a recurrence formula to provide another approximation. Several repetitions produce a sequence of solutions which, under suitable conditions, converge to the exact solution.

Consider again equation (3.27), which may be rewritten as

$$x_1 = 9 - x_2 + x_3 \qquad \text{(a)}$$
$$x_2 = (-1 + 2x_1 + x_3)/3 \qquad \text{(b)} \qquad\qquad (3.33)$$
$$x_3 = (7 - x_1 + x_2)/3 \qquad \text{(c)}$$

Assuming an initial solution of $x_1 = x_2 = x_3 = 0$, after 16 iterations

$$x_1 = 4.00469$$
$$x_2 = 3.00337$$
$$x_3 = 1.99956$$

The exact solution for these equations found previously was

$$x_1 = 4$$
$$x_2 = 3$$
$$x_3 = 2$$

The number of iterations used here was quite arbitrary. In practice, the iterations would be stopped using criteria based on the relative values at successive iterative steps. One such rule is to end when the modulus of the difference between the sum of the x values at the nth step and the $(n + 1)$th step are less than a given value, e.g.

$$\sum_i |x_i^{n+1} - x_i^n| < 0.00001$$

It is difficult to estimate the number of iterations required to achieve a particular accuracy. Convergence may in some cases be very quick and in other cases be too slow to be useful. It is possible to increase the chances of satisfactory convergence by rearranging the equations where possible so that the leading diagonal coefficient has an absolute value greater than others in its row. If the equations are badly arranged, it is possible for this method to produce results which diverge from the correct answer.

A further example of a numerical technique for the solution of a system of linear equations which may be written in tridiagonal matrix form is given in Chapter 8.

3.3 FINITE DIFFERENCE APPROXIMATIONS TO DIFFERENTIAL EQUATIONS

Before discussing finite difference methods it is useful to consider the way in which Taylor's series may be used to evaluate a function using a known value of the function and its derivatives close to the point of interest.

If it is assumed that the value of y at point B on the curve in Figure 3.3 is $f(t)$ and that point C has a value at B in the following way:

$$f(t + h) = f(t) + h^2 \frac{df(t)}{dt} + \frac{h^2}{2!} \frac{d^2 f(t)}{dt^2} + \frac{h^3}{3!} \frac{d^3 f(t)}{dt^3} \tag{3.34}$$

Similarly point A can be evaluated from point B

$$f(t - h) = f(t) - h \frac{df(t)}{dt} + \frac{h^2}{2!} \frac{d^2 f(t)}{dt^2} - \frac{h^3}{3!} \frac{d^3 f(t)}{dt} \tag{3.35}$$

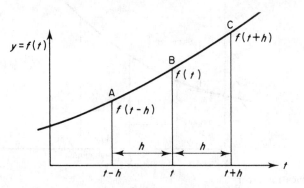

Figure 3.3. Taylor's series approximation of a differential equation

Assuming that $f(t)$ is continuously differential at B, the values calculated at C may be calculated to any degree of accuracy depending on the number of terms involved in the Taylor series. Although Taylor series are of no practical use for solving differential equations they are a useful basis for evaluating and comparing alternative numerical solution methods. For example, if a solution method agrees with the first three terms of the Taylor series then the truncation error is of order h^3, i.e. terms including the third derivative and all higher derivatives are ignored.

Consider the first-order differential equation

$$\frac{dy}{dt} = ky \tag{3.36}$$

which has a theoretical solution

$$y = A\,e^{kt} \tag{3.37}$$

where A is an arbitrary constant.

If A is assumed equal to one then in Figure 3.4 where $t = 0$, $y = 1$. Equation (3.35) may be used to evaluate y for any given value of t. Unfortunately, most differential equations are not amenable to analytic solution and may best be tackled by a numerical approach.

Equation (3.36) may be solved numerically given the initial value of $y = 1$ at $t = 0$. The numerical solution involves estimating successive values of y at points t_1, t_2, $t_3, \ldots, t_n$. The spacing of t_1, t_2, etc. is usually equidistant.

The Taylor series gives one method of obtaining a numerical solution. From equation (3.34), the first two terms of the expansion may be written as

$$f(t + h) \simeq f(t) + h\,\frac{df(t)}{dt} \tag{3.38}$$

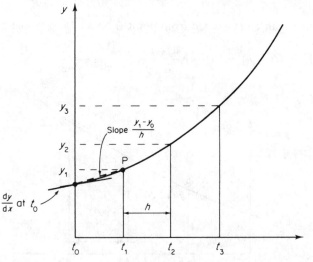

Figure 3.4. Truncation error

rearranging gives

$$\frac{d}{dt} f(t) = \frac{f(t+h) - f(t)}{h} \tag{3.39}$$

From Figure 3.3

$$f(t) = y_0 = 1 \qquad \text{evaluated at } t = 0 = t_0$$

$f(t+h)$ is the value of y at point P evaluated at t_1. Also

$$\frac{df(t)}{dt} = ky_0 \qquad \text{from equation (3.36)}$$

Thus equation (3.39) may be rewritten as

$$\frac{y_1 - y_0}{h} = ky_0$$

or

$$y_1 = y_0 + h/ky_0 \tag{3.40}$$

This is known as Euler's method, in which the derivative evaluated at the point y_0, t_0 is used to approximate the true value of y_1, as shown in Figure 3.5.

Having estimated y_1, this may be used to calculated y_2:

$$\frac{y_2 - y_1}{h} = ky_1$$

$$y_2 = y_1 + hky_1 \tag{3.41}$$

and in general

$$y_{n+1} = y_n + hky_n \tag{3.42}$$

Euler's method has truncation errors of order h^2 and unless h is kept very small the errors become large and the results inaccurate.

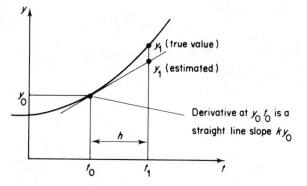

Figure 3.5. Simple Euler method for solving differential equations

3.3.1 Runge–Kutta Methods

The Runge–Kutta approach is based on the same principles as the Euler method but improves the accuracy of the result by reducing the truncation error when compared term by term with the Taylor expansion.

Consider again equation (3.36), which may be rewritten in the more general form

$$\frac{dy}{dt} = f(y) \tag{3.43}$$

In words, the derivative is equal to $f(y)$ and in Figure 3.5 ky_0 may be represented by $f(y_0)$ (the derivative evaluated at y_0, t_0).

The Runge–Kutta approach uses an estimate of the derivative evaluated at more than one point to estimate the new value of y_1.

A second-order method uses the average of slopes calculated at t_0, y_0 and at the point t_1, y_{1e} (the first estimate of y_1) to improve the approximation. This is shown graphically in Figure 3.6.

Now,

$$\frac{dy}{dt} \simeq \frac{y_1 - y_0}{h} = \frac{1}{2}(f(y_0) + f(y_{1e})) \tag{3.44}$$

and

$$y_1 = y_0 + \frac{h}{2}f(y_0) + \frac{h}{2}f(y_{1e}) \tag{3.45}$$

often equation (3.45) is expressed in the form

$$y_1 = y_0 + \tfrac{1}{2}(k_1 + k_2) \tag{3.46}$$

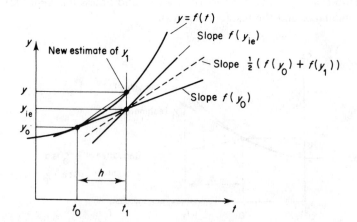

Figure 3.6. Runge–Kutta method for solving differential equations

where

$$k_1 = hf(y_0)$$
$$k_2 = hf(y_{1e})$$

now

$$y_{1e} = y_0 + hf(y_0) = y_0 + k_1 \qquad (3.47)$$

and

$$k_2 = hf(y_0 + k_1)$$

For a general function with the derivative a function of both y and t

$$\frac{dy}{dt} = f(t, y) \qquad (3.48)$$

The second-order Runge–Kutta can be written as

$$k_1 = hf(t_n, y_n)$$
$$k_2 = hf(t_n + h, y_n + k_1) \qquad (3.49)$$
$$y_{n+1} = y_n + \tfrac{1}{2}(k_1 + k_2)$$

Figure 3.7 shows graphically the new value of y_{n+1} based on the average of the slopes calculated at (t_n, y_n) and $(t_n + h, y_n + k_1)$

A common method in use is the fourth-order method, in which the average slope used to calculate y_{n+1} is based on a weighted average of the derivative evaluated at four separate points within the increment $t_n, t_n + h$.

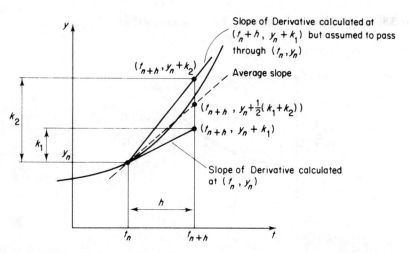

Figure 3.7. Second-order Runge–Kutta method

The fourth-order method for $dy/dt = f(t, y)$ may be written as

$$y_{n+1} = y_n + \tfrac{1}{6}(k_1 + 2k_2 + 2k_3 + k_4) \tag{3.50}$$

where

$$k_1 = hf(t_n, y_n)$$
$$k_2 = hg(t_n + \tfrac{1}{2}h, y_n + \tfrac{1}{2}k_1)$$
$$k_3 = hf(t_n + \tfrac{1}{2}h, y_n + \tfrac{1}{2}k_2)$$
$$k_4 = hg(t_n + h, y_n + k_3)$$

A graphical explanation of the procedure to evaluate $k_1, \ldots, k_4$ and hence obtain y_1 is shown in Figures 3.8–11.

Using the derivative evaluated at point 1, the value of k_1 can be calculated from $hf(t_n, y_n)$. Then k_2 is the increment of y corresponding to the increment h and the estimate of slope $f(t_n, y_n)$.

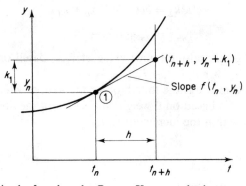

Figure 3.8. First step in the fourth-order Runge–Kutta method

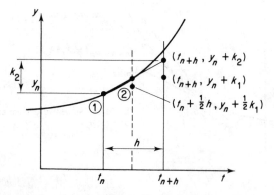

Figure 3.9. Second step in fourth-order Runge–Kutta method

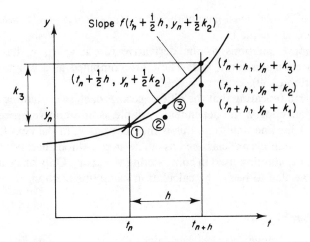

Figure 3.10. Third step in fourth-order Runge–Kutta method

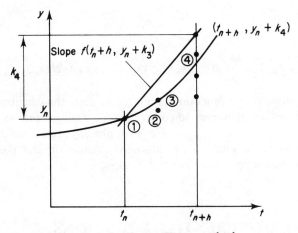

Figure 3.11. Fourth step in fourth-order Runge–Kutta method

Point 2 has coordinates $(t_n + h/2, y_n + (k_1/2))$ and the slope of the derivative of the function at point 2 is $f(t_n + h/2, y_n + (k_1/2))$. Then k_2 is the increment of y corresponding to the increment h and the estimate of slope $f(t_n + h/2, y_n + (k_1/2))$, assuming that the slope passes through point 1.

Point 3 has coordinates $(t_n + h/2, y_n + (\frac{1}{2}k_2))$ and the slope of the derivative of the function at point 3 is $f(t_n + h/2, y_n + (k_2/2))$. Then k_3 is the increment in y corresponding to the increment h and the estimate of the slope $f(t_n + (h/2), y_n + (k_2/2))$, assuming that the slope passes through point 1.

Point 4 has coordinates $(t_n + h, y_n + k_3)$ and the slope of the derivative of the function at point 4 is $f(t_n + h), (y_n + k_3)$. Then k_4 is the increment in y corresponding to the increment h and the estimate of the slope $f(t_n + h)$, $(y_n + k_3)$, assuming that the slope passes through point 1.

Thus four approximations to the derivative (dy/dt at t_n, y_n have been made producing four estimates for $y_n + 1$. These are combined as a weighted average as shown in equation (3.50).

It is again worth repeating that the difference approach to evaluating a function, by using incremental steps in the dependent variable, is an iterative procedure. Having evaluated $y_n + 1$ for one time step, this is then used as y_n in the next time step.

Runge–Kutta methods are classified as single-step methods because the only value of the approximate solution used in constructing y_{n+1} is y_n. Only one starting value of the function is needed to begin the calculation using this method.

3.3.2 Multi-Step Methods

Multi-step methods make use of earlier values like y_{n-1}, y_{n-2} etc. in order to reduce the number of times y has to be evaluated. Several methods are available, one of which is Milne's fourth-order method.

$$\frac{dy}{dx} = f(x, y)$$

$$y_{n+1} = y_{n-3} + \frac{4h}{3}\left(2f(x_n, y_n) - f(x_{n-1}, y_{n-1}) + 2f(x_{n-2}, y_{n-2})\right) \qquad (3.51)$$

More than one value of the function is required to start the calculation procedure, which is a disadvantage. However, to construct y_{n+1}, $f(x, y)$ need be evaluated only once since $f(x_{n-1}, y_{n-1})$, $f(x_{n-2}, y_{n-2})$ etc. have already been calculated.

Often Runge–Kutta may be used to start the calculation and then a multi-step approach used to speed up further computation.

3.4 PARTIAL DIFFERENTIAL EQUATIONS

Many partial differential equations describing pollution dispersion and decay in the environment cannot be solved analytically because of variable coefficients, boundary conditions or complexity. Finite difference methods are frequently used even when an analytical solution is available. Difference methods are approximate in the sense that an instantaneous derivative is approximated by a difference quotient over a small interval. However, they are not approximate in the sense of being crude estimates. Data used in environmental problems are subject to errors in measurement and also arithmetical calculations are subject to round-off errors, so that even analytical solutions give approximate numerical answers. Finite difference methods give solutions as accurate as the data warrants or as accurately as is necessary for the technical purposes for which the solution is required. In both cases the finite difference solution is as satisfactory as one calculated from an analytical formula.

The method may be illustrated using the one-dimensional convective diffusion equation for a non-conservative substance with a first-order decay rate:

$$\frac{\partial c}{\partial t} = E\frac{\partial^2 c}{\partial x^2} - U\frac{\partial c}{\partial x} - kc + L_a \tag{3.52}$$

Equation (3.52) may be used to describe an inland river with unidirectional flow, constant dispersion coefficient, constant cross-sectional area and constant velocity. Pollutant concentration will vary both with distance along the river and with time.

In the finite difference approach the area of integration is overlaid by a rectangular mesh and the approximate solution to the differential equation is found at the points of intersection of the mesh. In Figure 3.12, the x-axis is divided into increments of h and the time axis into increments of k; $c_{1,n}$ is the concentration value at the end of the ith distance interval and the nth time interval.

The partial derivatives in equation (3.52) are approximated using values at neighbouring mesh points to produce algebraic equations which may be solved directly or simultaneously depending on the form of the difference method employed.

3.4.1 Explicit Methods

The process is similar to that for ordinary differential equations in which the solution is obtained by stepping through time starting with a known concentration distribution. In the explicit method the unknown value $c_{i,n+1}$ is obtained using the three known values $c_{i-1,n}$, $c_{i,n}$ and $c_{i+1,n}$ to approximate the derivatives.

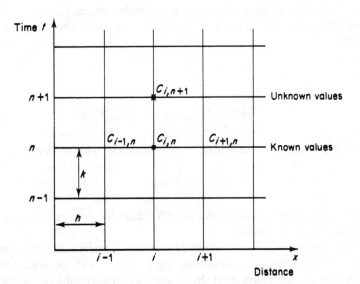

Figure 3.12. Finite difference mesh

Examining each term separately:

(1) The zero-order term kc can be represented as

$$kc = k\frac{(c_{i,n} + c_{i,n+1})}{2} \qquad (3.53)$$

and

$$L_a = \Delta L_i = \text{the increment of load discharged during}$$
$$\text{the time step at the point } i.$$

(2) The first-order term $\partial c/\partial t$ must be represented by a forward difference approximation in order to advance the time step of the computation.

$$\frac{\partial c}{\partial t} = \frac{c_{i,n+1} - c_{i,n}}{k} \qquad (3.54)$$

The term $U(\partial c/\partial x)$ may be represented by a backward difference approximation

$$U\frac{\partial c}{\partial x} = \frac{c_{i,n} - c_{i-1,n}}{h} \qquad (3.55)$$

This is necessary because any method which represents pure advection requires that

$$c_{i,n+1} = c_{i-1,n} \qquad (3.56)$$

In other words, a particle at the point $c_{i-1,n}$ must move a distance h in the time interval k to the point $c_{i,n+1}$ if the difference method is to provide a satisfactory solution. This can be illustrated using the simple advection equation

$$\frac{\partial c}{\partial x} = -U\frac{\partial c}{\partial x}$$

$$\frac{c_{i,n+1} - c_{i,n}}{k} = -U\frac{(c_{i,n} - c_{i-1,n})}{h} \qquad (3.57)$$

If the size of the mesh increments are chosen correctly so that

$$\frac{h}{K} = U$$

then it can be seen from equation (3.57) that the requirement specified by equation (3.56) is satisfied.

It can be shown that a forward difference or central difference approximation cannot satisfy this condition. Intuitively it can be seen that under conditions of pure convection the concentration at point i cannot depend on the concentration at $i + 1$.

(3) The dispersion term $E(\partial^2 c/\partial x^2)$ may be represented by a central difference approximation:

$$E\frac{\partial^2 c}{\partial x^2} = E\frac{(c_{i+1,n} - 2c_{i,n} + c_{i-1,n})}{h^2} \qquad (3.58)$$

This is derived from consideration of a Taylor series expansion about the point x_i. Taking the first three terms only for $c(x_i + h)$ and $c(x_i - h)$

$$c(x_i + h) = c(x_i) + \frac{h\,dc(x_i)}{dx} + \frac{h^2}{2}\frac{d^2c(x_i)}{dx^2}$$

$$c(x_i - h) = c(x_i) - \frac{h\,dc(x_i)}{dx} + \frac{h^2}{2}\frac{d^2c(x_i)}{dx^2}$$

Adding the above gives

$$c(x_i + h) + c(x_i - h) = 2c(x_i) + h^2\frac{d^2c(x_i)}{dx^2} \qquad (3.59)$$

rearranging the equation (3.59) gives

$$\frac{d^2c(x_i)}{dx^2} = \frac{1}{h^2}[c(x_i + h) - 2c(x_i) + c(x_i - h)] \qquad (3.60)$$

which is in the same form as equation (3.58) if $c(x_i + h)$ is written as c_{n+1}, etc.

It can be shown that a necessary condition for stability of solution with second derivatives of this form is

$$\frac{kE}{h^2} \le \tfrac{1}{2} \qquad (3.61)$$

The finite difference approximation to equation (3.52) can now be written as

$$\frac{c_{i,n+1} - c_{i,n}}{k} = E\frac{(c_{i+1,n} - 2c_{i,n} + c_{i-1,n})}{h^2}$$

$$-U\frac{(c_{i,n} - c_{i-1,n})}{h} - k\frac{(c_{i,n} + c_{i,n+1})}{2} + \Delta L_a \qquad (3.62)$$

Equation (3.62) may be rearranged to solve for the unknown value $c_{i,n+1}$ in terms of the known values of c at the nth time level. By solving equation (3.62) for each x mesh point at the $(n + 1)$th time level the new concentration profile can be computed. It is generally necessary and convenient to choose the spatial boundaries of the model in such a way that boundary concentrations are specified as constant for all time intervals.

It has been shown that the direct use of equation (3.6) can produce initial computational errors which may give misleading results. A two-step method has been suggested to overcome these problems. In this procedure the load of pollution is first convected downstream for one time step and then other processes such as dispersion and decay are carried out. In Figure 3.13 the concentrations are moved one mesh to the right starting at the right-hand end, i.e.

Step 1

$$c_{1+3,n} = k \quad \text{Boundary condition}$$

$$c_{1+2,n} = c_{i+1,n}$$

$$c_{i+1,n} = c_{i,n}$$

$$c_{i,n} = c_{i-1,n}$$

$$c_{i-1,n} = c_{i-2,n}$$

$$c_{i-2,n} = c_{i-2,n+1} \quad \text{Boundary condition}$$

Step 2

The new values of $c_{i,n}$ etc. can be used in equation (3.62) which no longer contains the advective term $U(dc/dx)$.

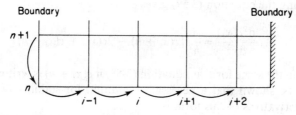

Figure 3.13. Explicit method of solving finite difference problems

Equation (3.52) may be written in a more general form as follows:

$$\frac{\partial AC}{\partial t} = \frac{\partial}{\partial x}\left(E \frac{\partial AC}{\partial x}\right) - \frac{\partial UAC}{\partial x} + S \tag{3.63}$$

This may be interpreted as an unsteady one-dimensional convective diffusion equation with cross-sectional area, velocity and dispersion coefficient varying with distance downstream. Source and decay terms are represented by the parameter S.

Using the notation in Figure 3.14 an explicit finite difference form of equation (3.63) may be expressed as

$$\frac{(AC)_{i,n+1} - (AC)_{i,n}}{k} = \frac{1}{h}\left[U_{i-1\frac{1}{2},n}(AC)_{i-1\frac{1}{2},n} - U_{i+1\frac{1}{2},n}(AC)_{i+1\frac{1}{2},n}\right]$$

$$+ \frac{1}{h}\left[E_{i+1/2,n}\left(\frac{\partial(AC)}{\partial x}\right)_{i+1/2,n} - E_{i-1/2,n}\left(\frac{\partial(AC)}{\partial x}\right)_{i-1/2,n}\right] + S \tag{3.64}$$

Equation (3.64) uses estimates of (AC) at the $\frac{1}{2}$ mesh points either side of position i to estimate the unknown value $(AC)_{i,n+1}$

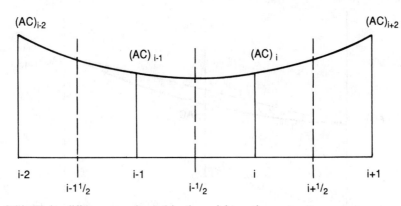

Figure 3.14. Finite difference mesh used in the quick routine

Leonard suggested that a stable accurate explicit (QUICK) procedure could be developed if a quadratic interpolation procedure was used to calculate the values of $AC_{i+1/2,n}$ and $AC_{i-1/2,n}$.

Hence $(AC)_{i+1/2,n} = \frac{1}{2}[(AC)_{i,n} + (AC)_{i+1,n}]$

$$- \frac{1}{8}[(AC)_{i-1,n} + (AC)_{i+1,n} - 2(AC)_{i,n}] \tag{3.65}$$

and

$$(AC)_{i-1/2,n} = \frac{1}{2}[(AC)_{i-1,n} + (AC)_{i,n}]$$

$$- \frac{1}{8}[(AC)_{i-2,n} + (AC)_{i,n} - 2(AC)_{i-1,n}] \tag{3.66}$$

As shown in Figure 3.15 the first terms of equation (3.65) and (3.66) represent linear interpolation and the second terms are proportional to the upstream weighted curvature of the concentration profile.

Both equations (3.65) and (3.66) assume the velocity flows from left to right. The gradient terms in equation (3.64) for the $i + \frac{1}{2}$ and $i - \frac{1}{2}$ mesh points may be expressed as

$$\left(\frac{\partial(AC)}{\partial x}\right)_{i+1/2,n} = \frac{(AC)_{i+1,n} - (AC)_{i,n}}{h} \tag{3.67}$$

and

$$\left(\frac{\partial(AC)}{\partial x}\right)_{i-1^{1}/2,n} = \frac{(AC)_{i,n} - (AC)_{i-1,n}}{h} \tag{3.68}$$

This is a consequence of the geometric properties of a parabola which ensure that the slope of the curve half-way between two points is equal to the slope of the chord joining these points.

Substitution of the estimates for $(AC)_{i+1/2,n}$ $(AC)_{i-1^{1}/2,n}$ $(\partial(AC)/\partial x)_{i+1/2,n}$ $(\partial(AC)/\partial x)_{i-1/2,n}$ into equation (3.64) produce an explicit equation with $(AC)_{i,n+1}$

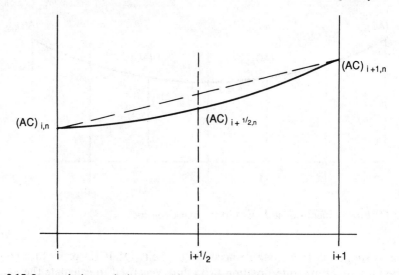

Figure 3.15 Interpolation technique used in the QUICK routine

expressed in terms of known (AC) values at the nth time step and at integer grid points in the x direction.

Inspection of equation (3.64) shows that concentrations at the $i - 2$, $i - 1$, i, and $i + 1$ mesh points are used to calculate the new C at the ith mesh point. It is also apparent that the second term in equation (3.65) may be rewritten as

$$- \tfrac{1}{8}[((AC)_{i+1,n} - (AC)_{i,n}) - ((AC)_{i,n} - (AC)_{i-1,n})]$$

It can be seen that this expression representing the curvature of the concentration profile at the $i + \tfrac{1}{2}$ mesh point is the difference between two gradients.

A similar expression may be used for the $i - \tfrac{1}{2}$ mesh point. An algorithm may be developed to calculate the new concentration at the $n + 1$ time step for all mesh points i based on the calculations of gradients between each adjacent pair of mesh points.

Let the gradient at $i + \tfrac{1}{2} =$

$$\frac{1}{h}((AC)_{i+1} - (AC)_i) = GR_r(i) \tag{3.69}$$

Hence the gradient at $i - \tfrac{1}{2} =$

$$\frac{1}{h}((AC)_i - (AC)_{i-1}) = GR_1 = GR_r(i - 1) \tag{3.70}$$

Let the curvature at $i + \tfrac{1}{2} =$

$$\frac{1}{h}(GR_r(i) - GR_r(i - 1)) = CRV(i) \tag{3.71}$$

Hence the curvature at $i - \frac{1}{2} =$

$$\frac{1}{h}(GR_r(i-1) - GR_r(i)) = CRV(i-1) \tag{3.72}$$

The following steps may be used in the QUICK calculation:

(1) Initialize all values of V, E, C, A, S at each mesh point.
(2) Calculate the gradients between all pairs of adjacent mesh points.
(3) Calculate the curvature associated with each $\frac{1}{2}$ mesh distance.
(4) Calculate the concentration of each $1\frac{1}{2}$ mesh point.

$$AC_{i+1/2,n} = \frac{1}{2}((AC)_{i,n} + (AC)_{i+1,n}) - \frac{h^2}{8}CRV(i) = AC^*(i) \tag{3.73}$$

Let $(AC)_{i-1/2,n} = AC^*(i-1)$

(5) Calculate for all i mesh values

$$(AC)_{i,n+1} = (AC)_i + \frac{k}{h}[U(i)AC^*(i) - U(i-1)AC^*(i-1)$$

$$- E(i-1)GR(i-1) + E(i)GR(i)] + ks \tag{3.74}$$

where s is the average source term.

The above steps assume that the velocities are positive from left to right in Figure 3.14. If the velocities are less than zero the curvature terms used to calculate $AC^*(i)$ and $AC^*(i-1)$ should be $CRV(i+1)$ and $CRV(i)$ respectively.

Leonard extended the QUICK method to describe unsteady flows which are primarily convective. In such cases the diffusion and source terms are relatively small. The QUICKEST scheme differs from QUICK in the definition of $(AC)_{i+1/2,n}$ and of $(\partial(AC)/\partial x)_{i+1/2,n}$

$$(AC)_{i+1/2,n} = \frac{1}{2}((AC)_{i,n} + (AC)_{i+1,n}) - \frac{h}{2}c_0(i)GR(i)$$

$$+ \frac{h^2}{2}[a(i) - \frac{1}{3}(1 + c_0^2(i)]CRV(i) \tag{3.75}$$

where

$$c_0(i) = (U_{i+1/2,n})k/h$$

and

$$a(i) = (E_{i+1/2,n})k/h^2$$

$$\left(\frac{\partial(AC)}{\partial x}\right)_{i+1/2,n} = GR(i) - \frac{h}{2}c_0(i)CRV(i)$$

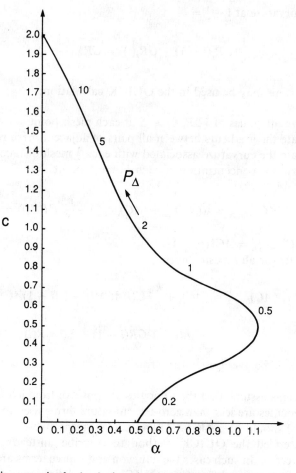

Figure 3.16. Stability range in the (α, c) plane for the QUICKEST method (note the extended range, in particular above $c = 1$). The nondiffusive limit $\alpha = 0$ ($P = \infty$) is unstable for $1 < c < 2$ and again for $c > 2$

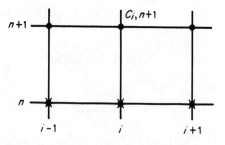

Figure 3.17. Implicit method of solving finite difference problems

Similar expressions may be obtained for $(AC)_{i-1/2,n}$ and

$$\left(\frac{\partial(AC)}{\partial x}\right)_{i-1/2,n}$$

The stability of the QUICK scheme depends on the choice of time and distance and mesh sizes. The two required conditions are

$$a + \frac{c_0}{4} \leq \tfrac{1}{2}$$

and

$$c_0^2 \leq 2a$$

For QUICKEST the necessary conditions are

$$\text{if } c_0 < \tfrac{1}{2} \qquad a \leq \frac{(3 - 2c_0)(1 - c_0^2)}{6(1 - 2c_0)}$$

$$\text{if } c_0 > \tfrac{1}{2} \qquad a \geq \frac{(3 - 2c_0)(c_0^2 - 1)}{6(2c_0 - 1)}$$

Also $a \geq 0$

These are necessary but not sufficient for stability. Leonard scanned a large range of a and c values to find the full stability range shown in Figure 3.16.

3.4.2 Implicit Methods

Implicit methods have been developed to overcome the restrictions of the explicit method, such as the ratio of the time and distance mesh increments.

In the implicit method (Figure 3.17), six points are used to approximate the derivatives at the i mesh distance leading to an equation involving three unknowns in the form:

$$\alpha c_{i-1,n+1} + \beta c_{i,n+1} + \gamma c_{i+1,n+1} = \delta(c_{i-1,n}, c_{i,n}, c_{i+1,n})$$

Equations are evaluated at each mesh point in the x direction and solved simultaneously at each time step. It is necessary for the boundary conditions to be specified so that the end equations only have two unknowns. A more comprehensive description of this approach is given in Chapter 8 on estuary models.

BIBLIOGRAPHY

Dresnack, R. and Dobbins, W. E. (1968). Numerical analysis of BOD and DO profiles. *Proc. ASCE.*, **SA5**, 789–807.

Foberg, C. E. (1970). *Introduction to Numerical Analysis.* (2nd edn) Addison-Wesley, New York.

Hosking, R. J., Joyce, D. C. and Turner, J. C. (1978) *First Steps in Numerical Analysis*, Hodder & Stoughton, London.

Monro, D. M. (1982). *FORTRAN 77*, Edward Arnold, Baltimore, USA.

Smith, G. D. (1978). *Numerical Solution of Partial Differential Equations: Finite Difference Methods*. Clarendon Press, Oxford.

Smith, J. M. (1977). *Mathematical Modelling and Digital Simulation for Engineers and Scientists*. Wiley, New York.

APPENDIX A

The following listing shows how a numerical routine may be coded and incorporated into a water quality model.

The subroutine QUICK is called in the main program at each time step. It needs to be supplied with the following information:

$U(I)$	velocity		$UN(I)$	new velocity
$Q(I)$	flow		$QN(I)$	new flow
$A(I)$	area		$AN(I)$	new area
$E(I)$	dispersion			
$P(I)$	sources			
$R(I)$	sinks			
$C(I)$	concentration			
DX	distance step			
DT	time step			
N	number of mesh points			

through COMMON or the ARGUMENT list.

The values of these variables are used to define the shape of the profile $AC(I)$ in terms of the gradient (GF) and the curvature (CVR). These in turn are used to calculate the effects on the rate of change of AC as PI_1 and PI_2.

In the final lines of calculation the new value of concentrations are derived from

   (i)   the effect of advection—line A in the listing
  (ii)   the effect of dispersions—line B in the listing
 (iii)   the effect of sources and sinks—line C in the listing

```
      SUBROUTINE QUICK
      COMMON/TNEW/FLN(101),VVELN(101),QN(101),UN(101),AN(101)
      DIMENSION E(101),P(101),R(101),C(101)
      DIMENSION AC(101),F(101),DP(101),AR(101),AN1(101),AC1(101)
      DIMENSION GF(101),UR(101),CR(101),UAR(101),AU(101),GAU(101)
      DIMENSION GGF(101),GCVR(101),GPI1(101),GP12(101),GCV(101)
      DIMENSION GCAU(101)
C
C
   10 Z = DT/DX
      NM1 = N − 1
      NM2 = N − 2
```

```
C
    8    DO 11 I = 1,N
         AC(I) = A(I)*C(I)
         DP(I) = E(I)*DT/DX/DX
C        AU(I) = A(I)*U(I)
   11    CONTINUE
C
C
         DO 12 I = 1,NM2
   12    GF(I) = (AC(I + 2) − AC(I + 1))/DX
         GF(NM1) = GF(NM2)
         GF(N) = GF(NM1)
         DO 13 I = 2,N
   13    CVR(I) = (GF(I) − GF(I − 1))/DX
         CVE(1) = CVR(2)
C
         DO 14 I = 2,N
         IF(I.EQ.N) GOTO 21
         UR(I) = (U(I + 1) + U(I))/2.0
         GOTO 22
   21    UR(I) = UR(I − 1)
   22    CONTINUE
         CR(I) = UR(I)*Z
         IF(UR(I).GE.O.O) CV(I) = CVR(I − 1)
         IF(UR(I).LT.O.O) CV(I) = CVR(I)
         IF(I.EQ.N) GOTO 23
C        UAR(I) = (AU(I + 1) + AU(I))/2.0 + DX*DX*GCAU(I)/8.0
         BA = 0.5*(AC(I) + AC(I + 1))
         GOTO 24
   23    BA = 0.5*(AC(I − 1) + AC(I))
   24    CONTINUE
         BB =  − DX/2.0*CR(I) + GF(I − 1)
         BC = DX*DX/2.0*(DP(I) − 1.0/3.0*(1.0 − CR(I)*CR(I)))*CV(I)
         PI1(I) = BA + BB + BC
         PI2(I) = GF(I − 1) − DX/2.0*CR(I)*CV(I)
   14    CONTINUE
         UR(1) = (U(2) + U(1))/2.0
         PI1(1) = (AC(2) + AC(1))/2.0 − 0.5*UR(1)*Z*(AC(2) − AC(1))
         PI2(1) = PI2(2)
C
C
C
         DO 101 I = 2,N
         F(I) = (Z*(UR(I − 1)*PI1(I − 1) − UR(I)*PI1(I)))
         F(I) = F(I) + Z*( − E(I − 1)*PI2(I − 1) + E(I)*PI2(I))
```

```
  201    CONTINUE
         F(I) = (AC(I) + F(I))/AN(I)
C        F(I) = F(I) + (P(I) − R(I))
C
  101    CONTINUE
         F(I) = C(1)
         F(N) = F(N − 1)
C
   66    RETURN
         END
```

Time Series

A. JAMES

4.1 INTRODUCTION

Monitoring of aquatic environments generally produces data in the form of repeated observations of some parameters at various sampling points. The repeated observation of a particular parameter at a particular point constitutes a time series. Such data possess special properties, such as persistence, which need to be considered in their analysis, both to avoid inappropriate statistical manipulation and to extract the full potential from the information.

Environmental time series may be collected in the following three forms:

(a) Continuous records from sensors like pH probes or current meters. These data need to be discretized into average values over time intervals so that they can be analysed. Care is needed in the choice of time interval.

(b) Discrete records from samples taken at particular points in time. The choice of sampling interval is crucial for this type of data.

(c) Cumulative records like daily rainfall where the parameter is already discretized and averaged by the sampling procedure.

In all cases the first step in examining a time series should be to plot the data on a time base. This allows the general properties like stationarity to emerge (see Section 4.2) and shows up the presence of outlying values which may need to be eliminated. A visual examination of the data is also useful in decisions on whether or not to transform the data. For example, if the variance seems to be proportional to the mean, then a log transform is indicated (see Figure 4.1).

Visual inspection of a time series is an essential preliminary step to the more detailed analysis outline in the following notes.

An Introduction to Water Quality Modelling. 2nd Edition. Edited by A. James

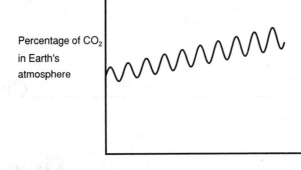

Percentage of CO_2 in Earth's atmosphere

Figure 4.1. Linear trend with increasing amplitude

4.2   CHARACTERISTICS OF TIME SERIES

A time series is a record of repeated observations made at a particular location. Each observation is a momentary summation of the effects of everything that is happening to the particular parameter being observed. It is assumed that the data discussed here are discrete or have been converted to a discrete form prior to analysis.

Time series may be in general divided into three groups:

(i)   those series that are largely periodic and may combine several distinct periodicities.

(ii) those series that contain some periodicity and a certain degree of randomness.

(iii) those series that appear to be characterized almost entirely by a random variation

Each type of series requires a somewhat different approach to extract information. For example once a type ii series has been analysed and its periodicities 'removed' or 'extracted' from the original record, a random time series may remain. Hence a given analysis scheme may uncover various types of time series that can then be studied individually.

A time series may be considered as the sum of three components: a trend component, a periodic or cyclic component and a random component. Any one or a combination of these components may exist in a particular series and it is interesting to consider the properties of each component separately to gain an insight into the analysis of time series.

4.2.1   Trend

The trend component is usually discernible only when the record is of long duration and may be defined as any frequency component whose period is longer than the record length. For example, a record of short length may exhibit an apparent trend that actually is a portion of a low-frequency periodicity. Hence care must be taken when identifying a long-term increase in the time-dependent parameter as a trend. It is possible to use least squares procedures for representing linear as well as higher-order polynomial trends.

Assume that the time series consists of observations X made at equally spaced time intervals $\Delta t = h$.

Also assume a linear relationship between x and t of the form

$$x(t) = A + Bt \tag{4.1}$$

where A and B are intercept and slope respectively for a straight line and x is the predicted value of x.

The difference between the observed X_i and predicted value of x_i may be written as

$$X_i(t) - x_i(t) = X_i(t) - (A + Bt_i) \tag{4.2}$$

and the sum of the squares of the deviations is given by

$$Q = \sum_{i=1}^{n} (X_i(t) - A - Bt_i)^2 \tag{4.3}$$

In order to minimize the sum of the squared deviations of the observed values from the predicted values, the first derivatives of (3) must be set to zero, i.e.

$$\frac{\partial Q}{\partial A} = \frac{\partial Q}{\partial B} = 0 \tag{4.4}$$

Solving for A and B gives

$$a = \bar{x} - b\bar{t} \tag{4.5}$$

$$b = \frac{\sum_{i=1}^{n} x_i t_i - n\bar{x}\bar{t}}{\sum_{i=1}^{n} t_i^2 - n\bar{t}^2} \tag{4.6}$$

where a and b are estimated for A and B for n observations and

$$\bar{x} = \frac{\sum_{i=1}^{n} x_i}{n}$$

$$\bar{t} = \frac{\sum_{i=1}^{n} t_i}{t}$$

and $\bar{x}$, $\bar{t}$ represent the mean values of x and t respectively.

In the prediction model $x_T(t)$, the trend component, can now be written as

$$x_T(t) = a + bt$$

$$= x - b\bar{t} + bt$$

$$x_T(t) = x + b(t - \bar{t}) \tag{4.7}$$

A similar least squares analysis may be used to fit non-linear trends by using a polynomial rather than a linear expression.

For example, let a polynomial of degree k be defined by

$$\hat{x}(t) = \sum_{j=0}^{k} b_j t^j$$

In a least squares fit, the coefficients b_j are chosen so as to minimize the expression.

$$Q(b) = \sum_{i=1}^{n} (x(t) - \hat{x}(t))^2$$

$$= \sum_{i=1}^{n} \left[x(t) - \sum_{j=0}^{k} b_j t^j \right]^2 \tag{4.8}$$

The degree k of the polynomial determines the number of coefficients b which must be estimated.

It is desirable to remove trend data in order to simplify subsequent formulae and calculations. If trends are not removed from time series data, to produce transformed data with zero mean, large distortions can occur in the later processing of correlation and spectral quantities. In particular trends in data can completely nullify the estimation of low frequent content. Figure 4.2 illustrates a case of linear trend removal.

4.2.2 Periodic Component

The periodic component of a time series may consist of one or more harmonics which combine to form a time-varying function whose waveform exactly repeats itself at regular intervals such that

$$x_F(t) = x(t \pm nT) \tag{4.9}$$

where T, the period, is the time interval required for one full fluctuation.

The harmonics can be described by the Fourier representation

$$x_F(t) = \sum_{k=1}^{m} (A_k \sin 2\pi f_k t + B_k \cos 2\pi f_k t) \tag{4.10}$$

where $x_F(t)$ is the estimate of the frequency component.
A_k, B_k are Fourier coefficients of the kth harmonic
f_k is the frequency (cycles per time unit)
k is an integer identifying the harmonic
m is the total number of harmonics considered non-negligible.

Frequency is related to the period of the first harmonic by $f_1 = 1/T$, and for other harmonics by $f_k = 1(T/k)$.

The frequencies observed in environmental phenomena may have periods anywhere from a fraction of a day to a year.

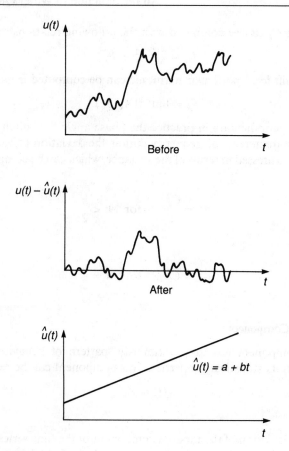

Figure 4.2. Linear trend removal

The best estimate of the Fourier coefficients A_k and B_k is given by (see Appendix A)

$$A_k = \frac{2}{n} \sum_{i=1}^{n} x_i \sin k2\pi f_1 t_i \tag{4.11}$$

$$B_k = \frac{2}{n} \sum_{i=1}^{n} x_i \cos k2\pi f_1 t_i \tag{4.12}$$

An alternative way to express the Fourier series for complex periodic data is as follows:

$$x_F(t) = \sum_{k=1}^{m} C_k \cos(k2\pi f_1 t - \theta_k)$$

in which all k harmonics need to be consecutive.

The amplitude C_k can be computed with the following relationship:

$$C_k = \sqrt{(A_k^2 + B_k^2)} \tag{4.13}$$

and the phase shift for a particular harmonic can be computed from

$$\theta_k = \tan^{-1}(A_k/B_k) \tag{4.14}$$

When analysing periodic data in practice the phase angle θ_k is often ignored.

An inherent characteristic of frequency data is the deviation about the mean. The deviation can be expressed in terms of the variance, which can be computed for a given harmonic with

$$\sigma_k^2 = \frac{C_k^2}{2} \quad \text{for} \quad k < \frac{n}{2}$$

or

$$\sigma_k^2 = C_k^2 \quad \text{for} \quad k = \frac{n}{2}$$

4.2.3 Random Component

The random component has no deterministic pattern of behaviour and may be predicted only in its statistical properties. This component can be represented by

$$x_R(t) = Z_n \sigma_k \tag{4.15}$$

where

σ_R = standard deviation of the random component of the time series.
Z_n = standardized normal variate with mean 0 and standard deviation unity.

Random variates with distributions other than normal may be substituted in (4.15); however, for most environmental phenomena of a random nature normal distributions provide a satisfactory representation.

A time series may be represented by a relationship which combines all three components—trend, frequency, random.

$$x(t) = x_T(t) + x_F(t) + x_R(t)$$

or

$$x(t) = \bar{x} + b(t - \bar{t}) + \sum_{k=1}^{m} C_k \cos(k 2\pi f t - \theta_k) + Z_n \sigma_R \tag{4.16}$$

The relative prominence of each component varies with the particular time series being considered. In general time series of environmental derivation fall into one of the following four categories:

(1) Time series that are composed of some periodicity, a certain degree of randomness, plus a mean with a time trend. Series of this type might be

observed in cases where stream water quality is monitored over a relatively long period of time in an area experiencing industrial development.

(2) Time series that are largely periodic and may include several distinct frequencies. Stream water temperature and tidal behaviour generally result in time series of this type.

(3) Time series that are composed of some periodicity and some degree of randomness. An example of this type can be found in the time records of dissolved oxygen in a river or estuary.

(4) Time series that appear to be characterized almost entirely by random variation. Over a relatively short time the average daily sewage flow to a waste treatment plant might yield this type of time series.

The above characterization depends not only on length of record but also on the particular statistic of the parameter of interest which is used. For example, although the average daily sewage flow may give a time series of type 4, the hourly flow may be in category 2 or 3.

4.3 ANALYSIS OF TIME SERIES

Information contained in time series may be extracted by a filtering procedure which splits the series into individual components. Such information may provide an insight into the cause–effect relationships influencing the parameter of concern.

The filtering procedure shown in Figure 4.2 removes the trend component, leaving the frequency and random components, then removes the frequency, leaving only the random component.

Variance is of significant interest in time series analysis. If all the observations in the time record are treated statistically a mean and variance can be computed; the latter is referred to as the total variance σ^2.

All three components contribute to the total variance

$$\sigma^2 = \sigma_T^2 + \sigma_F^2 + \sigma_R^2 \tag{4.17}$$

where

$$\sigma_T^2, \sigma_F^2, \sigma_R^2,$$

are the variances of trend, frequency and random components respectively.

The variance of each component can be isolated using the following procedure (see Figure 4.2).

First the total variance can be estimated from

$$\sigma^2 = \sum_{i=1}^{n} \frac{(x_i - \bar{x})^2}{n - 1} \tag{4.18}$$

The trend component is then removed and the resulting data analysed with (4.18) to obtain the variance contributed by the frequency and random components.

Removal of the frequency components from the observations leaves the residual data which will have a variance of σ_R^2.

The random component will only be truly random if all the frequency components are removed. This may be checked by plotting the residual data $x_R(t)$ on arithmetic probability paper, a straight line plot reveals normal distribution. If the residual does not plot as a straight line there is a possibility that all the frequency content has not been removed from the data.

4.3.1 Power Spectrum Analysis

For a given set of time series data it may not be immediately apparent which are the main frequencies causing the record to vary. Power spectrum analysis (or spectral analysis) enables the various components of the series to be identified by estimating the variance of each component. See Figure 4.3.

4.3.2 Autocovariance

In most series describing environmental behaviour the observations x_i are correlated with neighbouring observations x_{i-1} and x_{i-2} and the measure of such correlation is called the autocovariance function (see Figure 4.4).

$$R_{(\tau)} = \sum_{i=1}^{n-\tau} \frac{(x_i - \bar{x})(x_{i+\tau} - \bar{x})}{n - \tau} \qquad (4.19)$$

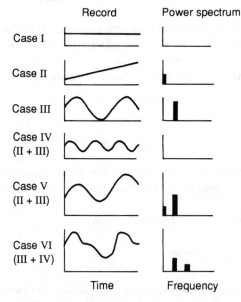

Figure 4.3. Power spectrum analysis

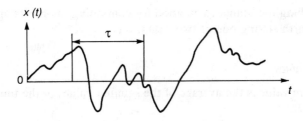

Figure 4.4. Autocorrelation measurement

where

$$x_i, x_{i+\tau} = \text{dependent variable at } i \text{ and } i + \tau$$
$$\tau = \text{time lag} = 0, 1, 2, \ldots, m.$$

It is often convenient to normalize $R(\tau)$ by dividing the variance of the series, which is simply the autocovariance with zero time lag $R(0)$. The result is the normalized autocorrelation function $R(\tau)_0$ which is dimensionless equal to unity at zero lag. See Figure 4.5.

Plot (a) in Figure 4.5 is typical of a random series in which the autocorrelation function falls to zero after a certain time lag and remains there.

Plot (b) indicates that the series has some periodicity. If the function remains positive it indicates that a persistence exists. A negative function indicates that high values will be followed by low values at some time τ later. No correlation or persistence is indicated when the function is zero.

Although the autocorrelation functions give an indication of periodicity the analysis may be taken a step further by the construction of power spectra plots which are plots of power spectral density versus frequency. These plots may be constructed either by the 'standard' method based on computing the power spectral density function, or by a finite range fast Fourier transform (FFT) of the original data, which is more efficient in terms of data computation.

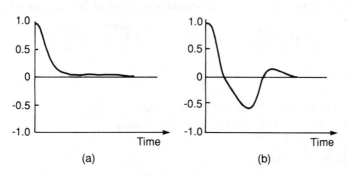

Figure 4.5. Autocorrelation functions (autocorrelogram)

Before describing the 'standard' method for computing the power spectral density function it is worth stating one or two definitions.

Mean Square Values

The mean square value is the average of the squared values of the time series

$$\Psi x^2 = \lim_{T \to \infty} \frac{1}{T} \int_0^T x^2(t)\, dt \tag{4.20}$$

where T is the total time span of the series.

This value may be related to the mean and variance of the time series by

$$\sigma x^2 = \Psi x^2 - \overline{x^2} \tag{4.21}$$

Now assume that the time series has been filtered, leaving a series with a frequency range between f and $f + \Delta f$.

$$\Psi x^2(f, \Delta f) = \lim_{T \to \infty} \frac{1}{T} \int_0^T x^2(t, f, \Delta f)\, dt \tag{4.22}$$

where $x(t, f, \Delta f)$ is that portion of $x(t)$ in the frequency range from f to $f + \Delta f$.

Then for small Δf a power spectrum density function can be defined as $G_x(f)$ such that

$$\Psi x^2(f, \Delta f) \simeq G_x(f)\Delta f \tag{4.23}$$

Now an important property of $G_x(f)$ is that it is related to the autocovariance function by a Fourier transform

$$G_x(f) = 2 \int_{-\infty}^{\infty} R_x(\tau)e^{-i2\pi f\tau}\, d\tau = 4 \int_0^{\infty} R_x(\tau) \cos 2\pi\tau\, d\tau \tag{4.24}$$

The second term exists because $R_x(\tau)$ is an even function of τ

Power Spectra Estimates from Correlation Estimates

A discrete approximation to the theroetical relation of (4.24) may be written as

$$\hat{G}_{x\tau} = \frac{k}{m}\left(R_0 + R_m \cos \tau\pi + 2 \sum_{i=1}^{m} R_i \cos \frac{i\tau\pi}{m} \right) \tag{4.25}$$

for $\tau = 0, 1, 2, m$

where $G_{x\tau}$ is the Fourier cosine transform at lag τ,

$R_0, R_m, R_i =$ autocovariances at lags $\tau, 0, m, i$

$$k = \begin{cases} 1 \text{ for } \tau = 1, 2, m \\ \tfrac{1}{2} \text{ for } \tau = 0 \text{ and } m \end{cases}$$

$i =$ lag number, having values between 1 and $m - 1$

From a statistical point of view this computation does not provide the best estimate of the spectrum. Various approaches have been taken to 'smooth' the estimates given by (4.25). All these approaches attempt to smooth out rapid fluctuations in the underlying 'true' spectrum. The smoothing procedure used is called 'Hanning' (from Julius Van Hann) and the smoothed estimates are given by

$$U_0 = \tfrac{1}{2}(G_0 + G_1)$$

$$U_\tau = \tfrac{1}{4}(G_{\tau-1} + \tfrac{1}{2}G_\tau + \tfrac{1}{4}G_{\tau+1}) \qquad 1 \le \tau \le m-1$$

$$U_m = \tfrac{1}{2}(G_{m-1} + G_m) \tag{4.26}$$

The dimensions of the spectral estimates computed by (4.26) are in variance per cycle per time. When the estimates are plotted against frequency (in cycles per time) the abscissa, the area under the spectrum, is the total variance of the record. If h is the interval between values of the series, then the dimensions of the abscissa are in cycles per $2mh$. Thus if $h = 1$ day and m (the maximum number of lags) is 30, the abscissa is labelled as cycles per 60 days and has 30 discrete values running from 0 to 30. The highest frequency of estimate (or shortest period) is given by

$$f_N = h/2 \tag{4.27}$$

in which f_N is the folding or Nyquist frequency. For $h = 1$ day $f_N = 1$ cycle per 2 days.

The power spectral estimates distribute the total variance of the record over a range of frequencies from zero (long-term trend) to a maximum of $h/2$. Any dominant periodicity will appear as peaks in the spectrum and the variance over the given frequency band indicates the contribution to the total variance of phenomena included within the given wave band. It should be noted that this computation does not give any information on the phase angle of a dominant frequency, only on the magnitude of the periodicity.

4.3.3 Aliasing

The designs of both the sampling interval and the subsequent statistical or spectral analysis require consideration of aliasing, which is defined graphically in Figure 4.6. Aliasing results from high-frequency events that add variance to the record but are not 'seen' by the particular sampling interval. Figure 4.6 shows how the variance from this type of event is folded into the record and reappears at a lower frequency. Any cyclic event that occurs at a period less than twice the sampling interval will result in aliasing. Where the period equals twice the sampling interval the event will never be seen. (The same is true of any event whose period (P) is related to the sampling interval h by $P = 2h/n$, in which n is a positive integer. In the design of sampling programs those cases in which $n > 1$ are not normally considered.) Only when the period exceeds twice the sampling interval can the event be measured.

From a practical standpoint it has been found that a sampling interval of no more than one third of the length of the shortest significant period is recommended. Thus to resolve a 12 hour period a sampling interval of no more than 4 hours should be used.

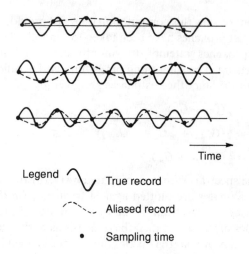

Time

Legend — True record

— Aliased record

• Sampling time

Figure 4.6. The influence of sampling frequency on the apparent effluent pattern

The longest period resolved is determined by the number of lags in the computation and by the sampling interval. Too many lags used in analysis will reduce the resolution of the spectral estimates, whereas too few will reduce the resolution of the spectral components.

For environmental time series an optimum number of lags to use is approximately 10% of the total number of observations in the length of record analysed. It is also recommended that the length of record to be analysed be at least 10 times as long as the longest significant period to be resolved.

4.3.4 Significance of the Harmonic Components

The height of the peaks of the variance spectra indicate the relative importance of the particular harmonic in contributing to the total variance of the record. Tests of significance of harmonic components are available, one of which is described below.

The value of the cumulative power spectrum at any frequency f can be obtained by integrating the spectrum form 0 to f_i, and the normalized cumulative power spectrum may be written as

$$C(f_i) = \frac{\sum_{i=1}^{n} G_\tau}{\sum_{f_i=1}^{0.5} G_\tau} \tag{4.28}$$

Equation (4.28) is a straight line for completely random data.

Confidence interals of $1.36/q$ can be placed on this line, where $q = (n-2)/2$ for an even number of lags and $(n-1)/2$ for an odd number. Deviation of the cumulative spectrum from a straight line will suggest that significant harmonics are included in this series.

Figure 4.7 shows two cumulative spectra: (a) shows a significant deviation from random data at a frequency of about 0.02 cycles/week and (b) shows a spectra with no significant deviation from random data.

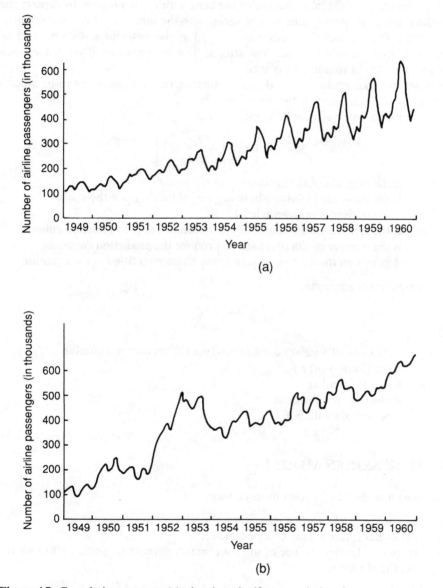

Figure 4.7. Cumulative spectra: (a) showing significant variation from random data; (b) showing no such significant variation

The procedure for analysing time series is therefore iterative. Trend and periodic components are identified and removed and the residuals are tested for further periodic components until all that remains is a completely random component.

4.3.5 Synthesis

Having identified all the components of the time series it is possible to reverse the procedure and generate a synthetic time series with the same statistical properties as the original. These synthetic traces may be used as the basis for a wide number of engineering uses including future low stream flow estimates or dissolved oxygen residual estimates in stream and ponds.

In order to make predictions with the synthetic series it is necessary to be able to construct confidence intervals for the values obtained.

The confidence interval may be constructed as

$$Y \pm t_\alpha \sqrt{\left[\sigma_R^2 \left(1 + \frac{1}{n} + A \right) \right]} \qquad (4.29)$$

where Y is the estimate of the variable from the equation;
t_α is the two-tailed t value whose degrees of freedom are those of σ^2;
α is the derived significance level;
σ_R^2 is the residual mean square *or* error variance in the prediction equation;
n is the number of samples used to produce the prediction equation;
A depends on the number of significant harmonics fitted in the equation.

For one significant harmonic

$$A = \frac{(x - \bar{x})^2}{\sum (x - \bar{x})^2}$$

where $\bar{x}$ is mean of all x values used to produce the prediction equation
x is $\cos [2\pi(n - m)/p]$
n is sample number
m is position of maximum
p is period of oscillation

4.4 TIME SERIES MODEL

Time series models may be used in three ways:

(a) to uncover and describe significant periodic behaviour;
(b) to predict future values of a parameter;
(c) to assess the significance of an abnormal treatment or perturbation on the predicted values.

To achieve (b) or (c) it is necessary to have some idea of whether new observed values differ significantly from prediction.

An additional test for the significance of deviation from predicted behaviour is based on the residual values before and after prediction.

If the model is appropriate after the perturbation has occurred, the residuals will be random. Therefore, if post-perturbation residuals are distributed with the same variance as those beforehand, no significance effect was caused by the perturbation.

The difference in residuals may be tested by comparing

$$\sum_{i=1}^{m} \frac{r_i^2}{\sigma_R^2} \quad \text{with} \quad \chi_n^2$$

where r_i = residual value after perturbation;
σ_R^2 = error variance for the prediction equation;
n = number of degrees of freedom;
χ_n^2 = chi-squared value for n degrees of freedom.

APPENDIX A

Fourier Series
A periodic function $f(x)$ with period 2π can be expanded in a series of sines and cosines.

$$f(x) = c_0 + c_1 \sin(x + \alpha_1) + c_2 \sin(2x + \alpha_2) + c_3 \sin(3x + \alpha_3) + \cdots \quad (1)$$

where the cs and αs are constants.

The terms $c_1 \sin(x + \alpha_1)$, $c_2 \sin(2x + \alpha_2)$ etc. are called the first and second harmonic respectively, and the process of finding them is called harmonic analysis.

c_0 does not affect the shape of the graph. This term may be removed by raising or lowering the x axis. It is in fact the mean value of the function over a period.

Expanding each term in (1),

$$f(x) = c_0 + c_1(\sin x \cos \alpha_1 + \cos x \sin \alpha_1)$$
$$+ c_2(\sin 2x \cos \alpha_2 + \cos 2x \sin \alpha_2)$$
$$+ c_3(\sin 3x \cos \alpha_3 + \cos 3x \sin \alpha_3) + \cdots$$
$$= c_0 + c_1 \sin \alpha_1 \cos x + c_2 \sin \alpha_2 \cos 2x$$
$$+ c_3 + \sin \alpha_3 \cos 3x + \cdots$$
$$+ c_1 \cos \alpha_1 \sin x + c_2 \cos \alpha_2 \sin 2x + c_3 \cos \alpha_3 \sin 3x + \cdots$$

let

$$\left.\begin{array}{l} a_n = c_n \sin \alpha_n \\ b_n = c_n \cos \alpha_n \end{array}\right\} \text{the } a_n \text{ and } b_n \text{ are therefore constants.}$$

Then

$$f(x) = a_0 + a_1 \cos x + a_2 \cos 2x + a_3 \cos 3x + \cdots$$
$$+ b_1 \sin x + b_2 \sin 2x + b_3 \sin 3x + \cdots \quad (2)$$

Equation (1) is usually the form required in practice. However, it is easier to calculate the coefficients in equation (2) and then transform into equation (1) using the relationship

$$c_n = \sqrt{(a_n^2 + b_n^2)} \quad \text{and} \quad a_n = \tan^{-1}(a_n/b_n) \tag{3}$$

Calculation of Coefficients a and b

The following standard integrals are used in the analysis

$$\int_{\pi}^{\pi} \cos nx \, dx = \int_{-\pi}^{\pi} \sin nx \, dx = 0 \tag{4}$$

$$\int_{-\pi}^{\pi} \cos mx \cos nx \, dx = \int_{-\pi}^{\pi} \sin mx \sin nx \, dx = \begin{cases} 0 \text{ if } m \neq n \\ \pi \text{ if } m = n \end{cases} \tag{5}$$

$$\int_{-\pi}^{\pi} \sin mx \cos nx \, dx = 0 \tag{6}$$

where m and n are integers other than zero.

To find a_0

Integrate both sides of (2) between $-\pi$ and π and use (4), giving

$$\int_{-\pi}^{\pi} f(x) \, dx = \int_{-\pi}^{\pi} \alpha_0 \, dx = a_0 2\pi$$

$$\therefore a_0 = \frac{1}{2\pi} \int_{-\pi}^{\pi} f(x) \, dx \tag{7}$$

$$= \text{mean value of } f(x) \text{ in range } (-\pi, \pi)$$

To find a_n

For $n \neq 0$

Multiply both sides of (2) by $\cos nx$ and integrate from $-\pi$ to π.
  Using (5) and (6) all the integrals on the right vanish except

$$\int_{-\pi}^{\pi} a_n \cos^2 nx \, dx \qquad \text{which equals} \quad a_n \pi$$

Thus

$$\int_{-\pi}^{\pi} f(x) \cos nx \, dx = a_n \pi \tag{8}$$

Hence

$$a_n = \frac{1}{\pi} \int_{-\pi}^{\pi} f(x) \cos nx \, dx$$

$$= \text{twice the mean value of } f(x) \cos nx \text{ in the range } -\pi \text{ to } \pi$$

To Find b_n

Multiply both sides of (2) by sin nx and integrate from $-\pi$ to π.
 Using (4), (5) and (6) all the integrals on the right vanish except

$$\int_{-\pi}^{\pi} b_n \sin^2 nx \, dx, \text{ which equals } b_n\pi$$

Thus

$$\int_{-\pi}^{\pi} f(x) \sin nx \, dx = b_n\pi$$

Hence

$$b_n = \frac{1}{\pi} \int_{-\pi}^{\pi} f(x) \sin nx \, dx \tag{9}$$

$$= \text{twice the mean value of } f(x) \, nx \text{ in the range } (-\pi, \pi)$$

If a_0 is calculated from the general formula using equation (8) with $n = 0$, the value obtained is

$$a_0 = \frac{1}{\pi} \int_{-\pi}^{\pi} f(x) \cos 0x \, dx = \frac{1}{\pi} \int_{-\pi}^{\pi} f(x) \, dx$$

which is twice the value obtained from (4).
 Usually the series is written using $\frac{1}{2}a_0$, a_0 having been calculated from the general formula. Then

$$f(x) = \tfrac{1}{2}a_0 + a_1 \cos x + a_2 \cos 2x + a_3 \cos 3x + \cdots$$

$$+ b_1 \sin x + b_2 \sin 2x + b_3 \sin 3x + \cdots \tag{10}$$

where

$$a_n = \frac{1}{\pi} \int_{-\pi}^{\pi} f(x) \cos nx \, dx$$

$$b_n = \frac{1}{\pi} \int_{-\pi}^{\pi} f(x) \sin nx \, dx \left.\vphantom{\int_{-\pi}^{\pi}}\right\} \tag{11}$$

This is known as the *Fourier series* for $f(x)$ of period 2π.

Expansion of a Periodic Function $f(x)$ of Period 1

The function $f(x)$ of period 1 repeats itself when x increases by 1, i.e. $f(x + 1) = f(x)$.
 By writing $\zeta = 2\pi x/1$ in which ζ increases by 2π when x increases by 1, then the function is now a periodic function of ζ with period 2π.
 The function can be expanded in the form:

$$\tfrac{1}{2}a_0 + \sum_{n=1}^{\infty} (a_n \cos n\zeta + b_n \sin n\zeta) \tag{12}$$

Hence

$$f(x) = \tfrac{1}{2}a_0 + \sum_{n=1}^{\infty} \left(a_n \cos n\frac{2\pi x}{1} + b_n \sin n\frac{2\pi x}{1} \right)$$

From above $x = \zeta 1/2\pi$ and $f(x) = f(\zeta 1)/2\pi = F(\zeta)$, and

$$a_n = \frac{1}{\pi} \int_0^{2\pi} f(\xi) \cos n\xi \, d\xi = \frac{2}{1} \int_0^1 f(x) \cos n\frac{2\pi x}{1} \, dx$$

$\qquad$ = twice the mean value of $f(x) \cos (n \, 2\pi x/1)$ in the range $(0, 1)$

$$b_n = \frac{1}{\pi} \int_0^{2\pi} F(\zeta) \sin n\zeta \, d\zeta = \frac{2}{1} \int_0^1 f(x) \sin n\frac{2\pi x}{1} \, dx$$

$\qquad$ = twice the mean value of $f(x) \sin (n \, 2\pi x/1)$ in the range $(0, 1)$.

Harmonic Analysis from Tabulated Values of a Function or from a Graph

The function to be analysed may be in the form of a discrete set of values rather than by a formula.

In this case the coefficients a_n and b_n cannot be found by integral calculus. They may, however, be found using one of the numerical integration rules such as the trapezoidal rule.

In the following discussion the period is assumed to be 2π. (It was shown earlier that a change of variable can be used to effectively reduce the period to 2π.)

Let the range 0 to 2π be divided into n equal parts with the dividing points defined as $x_0, x_1, x_2 \cdots x_n$.

Where an arbitrary point

$$x_i = i(2\pi/n) \text{ (interval} = 2\pi/n)$$

Let the ordinates at these points be $y_0, y_1, y_2, \ldots, y_n$ using the trapezoidal rule over period 2π:

$$a_0 = \frac{1}{\pi} \int_0^{2\pi} y \, dx = \frac{1}{\pi} \sum_{i=1}^{n} y_i = \frac{2}{n} \sum_{i=1}^{n} y_i$$

$$a_k = \frac{1}{\pi} \int_0^{2\pi} y \cos kx \, dx = \frac{1}{\pi} \frac{2\pi}{n} \sum_{i=1}^{n} y_i \cos kx_i = \frac{2}{n} \sum_{i=0}^{n} y_i \cos kx_i \qquad (13)$$

$$b_k = \frac{1}{\pi} \int_0^{2\pi} y \sin kx \, dx = \frac{1}{\pi} \frac{2\pi}{n} \sum_{i=1}^{n} y_i \sin kx_i = \frac{2}{n} \sum_{i=1}^{n} y_i \sin kx_i$$

Chapter 5

Modelling the Fate of Pollutants in Natural Waters—Processes and Mechanisms

A. JAMES

5.1 INTRODUCTION

Water quality models attempt to simulate changes in the concentration of pollutants as they move through the environment. There are some pollutants that are sufficiently inert for their concentration to be regarded as unchanging, except by physical transport phenomena like advection and dispersion. These are referred to as conservative substances and are often useful as tracers in the calibration and validation of water quality models. The basic mechanisms involved in processes like advection and dispersion are discussed in Section 5.2 since this affects both conservative and non-conservative substances.

Superimposed upon these mass transfer mechanisms, for the majority of pollutants, are physical, chemical and biological processes which also cause changes in the concentration. As shown in Figure 5.1 the fate of pollutants is the resultant of interactions between mass transfer and kinetic processes. The main controlling factors are the nature of the pollutant, the nature of the environment and the method of discharge. A pollutant entering the environment becomes partitioned between a series of subsystems, within each of which its concentration may increase or decrease due to a wide variety of mechanisms. It is apparent that it is not possible to consider the hydrosphere in isolation when studying the fate of pollutants. Terrestrial systems, especially the soil, may play a significant role. Processes such as leaching have therefore been included.

An Introduction to Water Quality Modelling. 2nd Edition. Edited by A. James
© 1993 John Wiley & Sons Ltd.

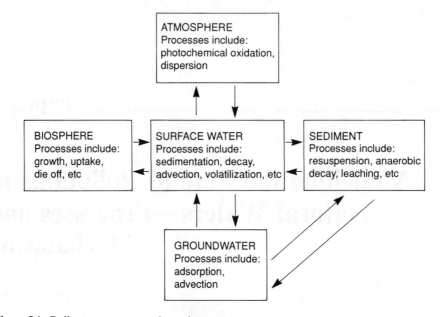

Figure 5.1. Pollutant movement through ecosystems

The following notes describe ways of representing the hydraulic, physical, chemical and biological processes, which determine pollutant concentration. Examples of the application of these in water quality models can be found in the later chapters on rivers, lakes, etc.

5.2 ADVECTION, DIFFUSION AND DISPERSION

There are two basic mechanisms that are responsible for the transport of dissolved and suspended solutes in natural waters, these are:

(a) *Advection* refers to transport due to the bulk movement of the water which contains the solute.

(b) *Diffusion* is the non-advective transport due to the migration of a solute in response to a concentration gradient. This can occur at the molecular level due to Brownian motion causing random movements of the solute molecule, or on a macroscopic scale due to turbulent eddies, and velocity shear.

These mechanisms are considered separately below:

5.2.1 Advection

The mechanism of advective transport is usually represented by the convective diffusion equation. Before considering the equation in detail it is useful to look briefly at its theoretical development.

The total mass of material entering into an element of space in a given time must equal the increase in mass within the space, in that time.

Figure 5.2 represents an element of volume, fixed in space relative to the Earth within a fluid stream. The principle of conservation of matter applies to the fluid itself and also to material dissolved in the fluid with density or suspended in the fluid. Assume initially that a fluid with density ρ flows through the box and that the fluid velocity has components u, v and w in the x, y and z directions. The flux of material passing Plane 1 is equal to the density multiplied by the velocity $= \rho u\, \delta y\, \delta z$.

Because the elemental volume is considered to be infinitesimally small it is possible to use Taylor's expansion to describe the flux passing Plane 2, i.e.

$$\left[\rho u + \frac{\partial(\rho u)}{\partial x}\right]\delta y\, \delta z$$

Hence the net decrease in mass due to flow in the x direction in the time δt is

$$\frac{\partial(\rho u)}{\partial x}\, \delta x\, \delta y\, \delta z\, \delta t$$

Similar expressions for the y and z direction may be obtained.

Assuming the initial mass in the element at time t is $\rho\, \delta x\, \delta y\, \delta z$, then the mass at time $t + \delta t$ can again be obtained from a Taylor expression as

$$\left[\rho + \frac{\partial\rho}{\partial t}\, \delta t\right]\delta x\, \delta y\, \delta z$$

Giving the rate of mass change within the element as

$$\frac{\partial\rho}{\partial t}\, \delta t\, \delta x\, \delta y\, \delta z$$

Equating the three mass flux terms with the rate of change gives

$$-\frac{\partial\rho}{\partial t}\, \delta t\, \delta x\, \delta y\, \delta z = \left[\frac{\partial(\rho u)}{\partial x} + \frac{\partial(\rho v)}{\partial y} + \frac{\partial(\rho w)}{\partial z}\right]\delta x\, \delta y\, \delta z\, \delta t$$

$$-\frac{\partial\rho}{\partial t} = \frac{\partial(\rho u)}{\partial x} + \frac{\partial(\rho v)}{\partial y} + \frac{\partial(\rho w)}{\partial z} \tag{5.1}$$

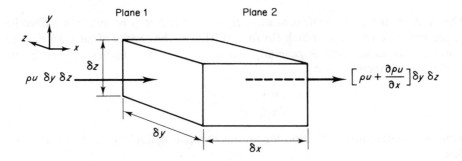

Figure 5.2. Conservation of mass

In equation (5.1) $\partial/\partial t$ is often called the local term which describes changes that would occur independent of the movement of the fluid particle. In effect, it is the rate of change that would occur for a motionless particle at a certain point.

$$-\frac{\partial\rho}{\partial t} = \rho\left(\frac{\partial u}{\partial x} + \frac{\partial v}{\partial y} + \frac{\partial w}{\partial z}\right) + \frac{u\partial\rho}{\partial x} + \frac{v\partial\rho}{\partial y} + \frac{w\partial\rho}{\partial z} \tag{5.2}$$

The terms $u(\partial/\partial x) + v(\partial/\partial y) + w(\partial/\partial z)$ are 'convective' terms which describe the rate of change due to the particle moving in a field where gradients of the property exist.

The combined effect of $\partial/\partial t + u(\partial/\partial x) + v(\partial/\partial y) + w(\partial/\partial z)$ is often referred to as the substantial or total derivative in fluid mechanics and represents the total rate of change of some property of the fluid experienced by a particular particle of fluid as it moves with velocity components u, v and w.

The total derivative may be expressed as

$$\frac{D}{Dt} = \frac{\partial}{\partial t} + u\frac{\partial}{\partial x} + v\frac{\partial}{\partial y} + w\frac{\partial}{\partial z} \tag{5.3}$$

An example may help to clarify the physical significance of these terms.

Assume an aeroplane is flying south (x direction) at a constant height. Throughout the area of flight the temperature is increasing by 1°C per day. Also for each 1000 miles of travel the temperature increases by 2°C. Thus the local effect experienced by a stationary body would be an increase in temperature of 1°C/day.

The total rate of increase experienced by the aircraft is

$$\frac{D}{Dt} = \frac{\partial T}{\partial t} + u\frac{\partial T}{\partial x} = 1 + \frac{u \times 2}{1000} \tag{5.4}$$

where T is the mean daily temperature. If the velocity u is 200 miles per hour, then the total rate of increase experienced by the plane is

$$\frac{D}{Dt} = 1 + 200 \times 24 \times \frac{2}{1000} = 1 + 9.6 = 10.6°C/day$$

Returning to equation (5.2), this may now be expressed as

$$\frac{D\rho}{Dt} + \rho\left(\frac{\partial u}{\partial x} + \frac{\partial v}{\partial y} + \frac{\partial w}{\partial z}\right) = 0 \tag{5.5}$$

This equation is known as the continuity equation which may be further modified for an incompressible fluid for which $D\rho/Dt = 0$. That is, the density of the fluid remains constant over time, which is generally true for normal river and estuary problems. Equation (5.5) then becomes

$$\frac{\partial u}{\partial x} + \frac{\partial v}{\partial y} + \frac{\partial w}{\partial z} = 0 \tag{5.6}$$

which is the continuity equation for an incompressible fluid in steady or unsteady flow.

Although changes in density may be considered insignificant over the depths encountered in rivers and estuaries, the concentrations of salts in solutions or solids in suspension may change rapidly over the distance involved.

Movement with the bulk water flow, advection, is not the only physical mechanism which must be considered, and random movements due to turbulence of the fluid or interchange of material between fast- and slow-moving layers of fluid must also be considered. These processes cause dispersion of material suspended in the liquid mass. This dispersion is relative to the bulk liquid flow and may be described mathematically in a form analogous to the molecular diffusion process.

5.3 MOLECULAR DIFFUSION AND FICK'S LAW

Assume that a conservative material is suspended or dissolved in a liquid.

Fick's first law states that the rate of mass transport of material or flux through a unit area of liquid by molecular diffusion is proportional to the concentration gradient of the material in the liquid.

$$\text{Diffusive mass flux } N = -D_m \frac{\partial C}{\partial x}$$

where D_m is the molecular diffusion coefficient or constant of proportionality (Figure 5.3). It is proportional to absolute temperature and also inversely proportional to the molecular weight of the diffusing phase and the viscosity of the dispersing phase. The negative sign is an indication that material flows from areas of high concentration to areas of low concentration due to the diffusive process.

Fick's second law may be obtained by again considering an elemental volume fixed in space in which the transport mechanism is pure diffusion (Figure 5.4).

By equating the time rate of change of mass within the element to the change caused by the diffusive flux in each of the three coordinate directions

$$\frac{\partial c}{\partial t} = -\frac{\partial}{\partial x} D_m \frac{\partial c}{\partial x} - \frac{\partial}{\partial y} D_m \frac{\partial c}{\partial y} - \frac{\partial}{\partial z} D_m \frac{\partial c}{\partial z}$$

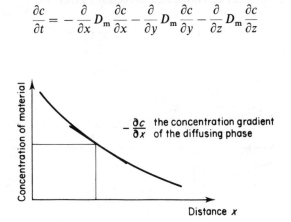

Figure 5.3. Fick's first law of diffusion

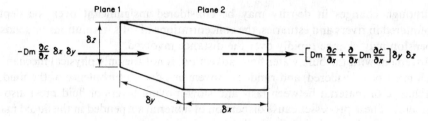

Figure 5.4. Fick's second law of diffusion

and assuming that the molecular diffusion coefficient D_m is constant in all coordinate directions

$$\frac{\partial c}{\partial t} = -D_m \left[\frac{\partial^2 c}{\partial x^2} + \frac{\partial^2 c}{\partial y^2} + \frac{\partial^2 c}{\partial z^2} \right] \tag{5.7}$$

Equation (5.7) describes the rate of change in concentration with respect to time of a substance subject only to the molecular diffusion process, and the use of the constant of proportionality D_m is restricted to this context.

The dispersion of effluents subject to large-scale processes of advection and turbulence may be described mathematically by an analogy with Fick's laws of diffusion.

Consider Figure 5.5, in which the elemental volume is fixed in a fluid medium. The conservation of mass equation of a substance introduced into the fluid can be obtained in a similar way to that used for the continuity equation.

Neglecting molecular diffusion relative to the large-scale mixing due to turbulence, the flux of material into the element across Plane 1 is $cu\ \delta y\ \delta z$, where u is the instantaneous velocity in the x direction. The net change in mass of material in the element from the flux in the x direction is $(\partial/\partial x)cu\ \delta x\ \delta z$.

Equating the time rate of change of mass in the element with the rate of change due to the flux in each of the three coordinate directions gives

$$\frac{\partial c}{\partial t} + \frac{\partial}{\partial x}(cu) + \frac{\partial}{\partial y}(cv) + \frac{\partial}{\partial z}(cw) = 0 \tag{5.8}$$

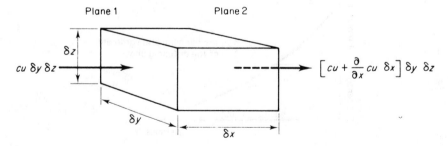

Figure 5.5. Conservation of mass of a solute

The values of concentration and velocity, although in theory instantaneous, are in practice measured over short periods of time. Thus the instantaneous velocity component u can be expressed as

$$u = \bar{u} + u'$$

where

$$\bar{u} \text{ is a time-averaged component} = \frac{1}{T} \int_{t}^{t+T} u \, dt \qquad (5.9)$$

and T is the period of measurement (Figure 5.6).
   Also

$$v = \bar{v} + v'$$

$$w = \bar{w} + w'$$

$$c = \bar{c} + c'$$

These expressions for instantaneous velocities can be substituted into equation (5.8) and each term averaged to give

$$\frac{\partial}{\partial t} (\bar{c} + c') + \frac{\partial}{\partial x} (\bar{c} + c')(\bar{u} + u') + \frac{\partial}{\partial y} (\bar{c} + c')(\bar{v} + v')$$

$$+ \frac{\partial}{\partial z} (\bar{c} + c')(\bar{w} + w') = 0 \qquad (5.10)$$

The terms in brackets may be expanded and all terms with only one prime will average to zero. For example, a term such as $c'u = 0$.
   Then

$$\frac{\partial \bar{c}}{\partial t} + \frac{\partial}{\partial x} (\overline{uc}) + \frac{\partial}{\partial y} (\overline{vc}) + \frac{\partial}{\partial z} (\overline{wc}) + \frac{\partial}{\partial x} (\overline{u'c'})$$

$$+ \frac{\partial}{\partial y} (\overline{v'c'}) + \frac{\partial}{\partial z} (\overline{w'c'}) = 0 \qquad (5.11)$$

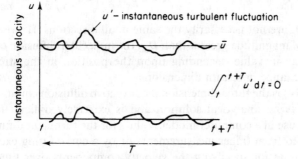

Figure 5.6. Velocity variations with time due to turbulence

The derivatives of products may be expanded in the form $u(dv/dx) + v(du/dx)$, giving

$$\frac{\partial \bar{c}}{\partial t} + \bar{u}\frac{\partial \bar{c}}{\partial x} + v\frac{\partial \bar{c}}{\partial y} + w\frac{\partial \bar{c}}{\partial z} + \bar{c}\left(\frac{\partial \bar{u}}{\partial x} + \frac{\partial \bar{v}}{\partial y} + \frac{\partial \bar{w}}{\partial z}\right)$$

$$+ \frac{\partial}{\partial x}\overline{(uc')} + \frac{\partial}{\partial y}\overline{(v'c')} + \frac{\partial}{\partial z}\overline{(w'c')} = 0 \qquad (5.12)$$

Using the continuity equation

$$\frac{\partial \bar{u}}{\partial x} + \frac{\partial \bar{v}}{\partial y} + \frac{\partial \bar{w}}{\partial z} = 0$$

Equation (5.12) simplifies to

$$\frac{\partial \bar{c}}{\partial t} + \bar{u}\frac{\partial \bar{c}}{\partial x} + v\frac{\partial \bar{c}}{\partial y} + w\frac{\partial \bar{c}}{\partial z} + \frac{\partial}{\partial x}\overline{(u'c')} + \frac{\partial}{\partial y}\overline{(v'c')} + \frac{\partial}{\partial z}\overline{(w'c')} = 0 \qquad (5.13)$$

The cross-product terms such as $u'c'$ represent the net convection of mass due to the turbulent fluctuations and by analogy with Fick's law of molecular diffusion they can be represented by an equivalent diffusive mass transport system in which the mass flux is proportional to the mean concentration gradient and the flux is in the direction of the mean concentration gradient. Hence

$$\overline{u'c'} = -D_x\frac{\partial \bar{c}}{\partial x}$$

$$\overline{v'c'} = -D_y\frac{\partial \bar{c}}{\partial y}$$

$$\overline{w'c'} = -D_z\frac{\partial \bar{c}}{\partial z}$$

where D_x, D_y and D_z are coefficients of turbulent diffusion. Equation (5.13) can now be rewritten omitting the time average bars as

$$\frac{\partial c}{\partial t} + u\frac{\partial c}{\partial x} + v\frac{\partial c}{\partial y} + w\frac{\partial c}{\partial z} + \frac{\partial}{\partial x}\left(D_x\frac{\partial c}{\partial x}\right)\frac{\partial}{\partial y}\left(D_y\frac{\partial c}{\partial y}\right)\frac{\partial}{\partial z}\left(D_z\frac{\partial c}{\partial z}\right) = 0 \qquad (5.14)$$

D_x, D_y and D_z are not necessarily the same in all directions. However, they will be several orders of magnitude greater than D_m, the molecular diffusion coefficient. They may also change in value depending upon the position in the stream or estuary because of the change in stream dimensions.

Equation (5.14) is the three-dimensional convective diffusion equation which in this general form has no analytical solution and is extremely difficult to approximate numerically for use in a computer model. Values for the turbulent diffusion coefficient must be obtained from tracer measurements or by a curve-fitting exercise and some way must be found for specifying the velocity components over time. This may be done by first solving the momentum equation and using the resulting velocities in the convective diffusion equation, or by calculating velocities from volume changes in the case of estuaries subject to tidal fluctuations. Whichever method is used, a knowledge of freshwater flow is required over the period of time being considered. For these reasons simplifying assumptions are usually made to reduce equation (5.14) to a form amenable for solution.

Equation (5.14) may be reduced to a one-dimensional form by taking average values of all parameters across the section of the stream or estuary and considering concentration only as a function of x and t.

For example,

$$U = \frac{1}{A} \int_A^0 \bar{u} \, dA$$

$$C = \frac{1}{A} \int_A^0 \bar{c} \, dA \tag{5.15}$$

where A is the cross-sectional area.

$$\frac{\partial c}{\partial y} \quad \text{and} \quad \frac{\partial c}{\partial z} = 0, \quad v = w = 0$$

and equation (5.14) reduces to

$$\frac{\partial c}{\partial t} + U \frac{\partial c}{\partial x} = \frac{1}{A} \frac{\partial}{\partial x} \left(EA \frac{\partial c}{\partial x} \right) \tag{5.16}$$

where E is now an effective longitudinal dispersion coefficient.

Equation (5.16) has been used for both rivers and estuaries and under certain assumptions may be simplified even further.

5.4 MASS TRANSFER

If the concentrations are different in two parts of a system, the system will tend to adjust itself to an equilibrium condition. The rate at which this adjustment takes place is proportional to the magnitude of the difference, which is called the *driving force* or *driving potential*, and to the interfacial surface area through which transfer occurs. The fluid or solid through which the chemical moves offers resistance to this adjustment

$$\text{mass flux} = \frac{(\text{area})(\text{driving force})}{\text{resistance}} \tag{5.17}$$

Resistance is usually described in terms of a *mass transfer coefficient*. An overall mass transfer coefficient is a lumped parameter that incorporates the effects of turbulence, chemical type, etc., and is typically a function of temperature.

A pond contains a chemical at uniform concentration C. The chemical can move across the surface of the pond between the water and the atmosphere. The surface area is A, the pond volume is V, and the ratio of volume to area is $d = V/A$ (d is sometimes called the average 'depth'). The material balance equation is

$$\frac{V \, dC}{dt} = - kA(C_w - C_a)$$

(5.18)

or

$$\frac{dC}{dt} = - k\frac{C}{d}$$

(5.19)

which has solution

$$C = C_0 \exp\left(- k\frac{t}{d}\right)$$

(5.20)

where C_w = chemical concentration in the water,
C_a = chemical concentration in the atmosphere above the pond surface, and
C_0 = chemical concentration in the water at time $t = 0$.
k = an overall mass transfer coefficient with units of L/T.

The direction of movement is determined by the sign of the quantity $(C_w - C_a)$. Say, for example, that a synthetic chemical has been spilled into the pond. Then $C_w > C_a$ and movement is from the water to the air.

If the chemical of interest is, say, oxygen and the water is not saturated, then $C_w < C_a$ and oxygen would transfer from the atmosphere into the water. In this case, the mass transfer coefficient is often called the reaeration coefficient. Equation (5.18) is often written specifically as a model for reaeration (oxygen transfer from the atmosphere to a water body):

$$V\frac{dc}{dt} = K_L A(c_s - c)$$

(5.21)

where c = dissolved oxygen concentration in the water, mg/L
c_s = saturation concentration of oxygen in the water, mg/L, and
K_L = interfacial reaeration coefficient (L/T)

This can be written in terms of a volumetric reaeration coefficient, K_a, as

$$\frac{dc}{dt} = K_a(c_s - c)$$

(5.22)

where $K_a = K_L A/V$ has units of T^{-1}. A commonly used form of this is written in terms of the oxygen deficit, $D = c_s - c$. Since $dD = dc_s - dc$,

$$\frac{dD}{dt} = - K_a D - \frac{dc_s}{dt}$$

(5.23)

If it assumed that the factors that affect the saturation concentration c_s (temperature, salinity, and pressure) are constant, then $dc_s/dt = 0$ and

$$\frac{dD}{dt} = - K_a D$$

(5.24)

To look at this phenomenon in more detail, imagine that a chemical moves by diffusion across a gas–liquid boundary that consists of two films, one of gas and one of liquid. When this system is at steady state, the rate of mass transfer across the gas film and the liquid film must be equal:

$$\text{Flux}_{\text{gas}} = -D_{\text{gas}}\left(\frac{dC}{dz}\right)_{\text{gas}} = -D_{\text{liquid}}\left(\frac{dC}{dz}\right)_{\text{liquid}} = \text{Flux}_{\text{liquid}} \tag{5.25}$$

The diffusivity, D, in the gas phase is different than in the liquid phase. Also, the thickness of the gas and liquid 'films' are different.

At the interface the gas and liquid phases must be in equilibrium, and this condition is defined by Henry's law: the mass of any gas that dissolves in a given volume of liquid, at a constant temperature, is directly proportional to the pressure that the gas exerts above the liquid. Henry's constant is the ratio of the partial pressure of the gaseous phase to the solubility of the chemical in the water phase.

$$p = Hc_s \tag{5.26}$$

where p = partial pressure of the chemical, mm Hg,
$\quad c_s$ = saturation concentration of the chemical in liquid, mg/l,
$\quad H$ = Henry's constant, mm Hg.

A dimensionless form is

$$H = \frac{16Mp}{Tc_s} \tag{5.27}$$

where T is temperature in K and M is the molecular weight in g/g-mole, and H has units of mg/l in the gas phase per mg/l in the water phase. This form is often used in water quality modelling. Care must be taken to make certain units are used consistently.

The fluxes in equation (5.25) can be written in terms of the bulk conditions in either the gas or the liquid phases:

gas phase
$$F = K_g(c_g = Hc_L) \tag{5.28}$$

liquid phase
$$F = K_L\left(\frac{cg}{H-c}\right) \tag{5.29}$$

where K_g is the overall mass transfer coefficient in terms of the gas phase, and K_L is the same coefficient in terms of the liquid phase. These can be expressed in terms of gas phase and liquid phase coefficients, as follows:

$$K_g = \frac{1}{k_g} + \frac{H}{k_L} \tag{5.30}$$

and

$$K_L = \frac{1}{k_L} + \frac{1}{Hk_g} \tag{5.31}$$

For chemicals of low solubility, H is large, and the flux rate will be controlled by the liquid-side resistance (and is insensitive to the value of k_g). As an approximation,

$$k_g = 3000 \sqrt{\frac{18}{\text{mol.wt}}} \tag{5.32}$$

and

$$k_L = 20 \sqrt{\frac{44}{\text{mol.wt}}} \tag{5.33}$$

If the oxygen transfer rate is unknown it can be used to estimate the mass transfer rate for other chemicals. The ratio below has been verified for chemicals that have the main resistance to mass transfer in the liquid film layer (as oxygen does).

$$K_c = K_L a \left(\frac{k_c}{k_{O_2}} \right) \tag{5.34}$$

where $K_L a$ is the overall mass transfer rate for oxygen that is estimated for field conditions. k_c and k_{O_2} are mass transfer rates for the chemical and oxygen, measured under similar laboratory conditions.

Diffusion

From the previous discussion (see Section 5.2) it has been demonstrated that Fick's law of mass transport says that the mass flux is proportional to the concentration gradient. The flux through a cross section of area A is

$$N_a = - D A \frac{\partial c}{\partial z}$$

where D has units of L^2/T.

If substance A diffuses, say from a gas phase into a liquid film, and reacts, the material balance equation is

$$\frac{\partial c}{\partial t} = - \frac{\partial}{\partial z} \left(- D_A \frac{\partial c}{\partial z} \right) - r_A$$

where r_A is the reaction rate for which an appropriate expression must be provided.

If substrate A is transported by advection as well as diffusion, the material balance equation is

$$\frac{\partial c}{\partial t} = - \frac{\partial}{\partial z} \left(- D_A \frac{\partial c}{\partial z} \right) + \frac{\partial}{\partial z} (uc) - r_A$$

Example: Suppose that substance A is transported across a gas-liquid interface and it reacts according to

$$A \xrightarrow{k_1} P_s \qquad\qquad r_A = - k_1 c_A - k_2 c_A c_B$$

$$A + bB \xrightarrow{k_2} Q_s \qquad\qquad r_B = - k_1 c_A c_B$$

The one-dimensional, unsteady-state material balance model for substance A is

$$\frac{\partial c}{\partial t} + D_A \frac{\partial^2 c_A}{\partial z^2} - k_1 c_A - k_2 c_A c_B$$

and for B is

$$\frac{\partial c}{\partial t} = D_B \frac{\partial^2 c_B}{\partial z^2} - b k_2 c_A c_B$$

At steady state

$$D_A \frac{\partial^2 c_A}{\partial z^2} - k_1 c_A - k_2 c_A c_B = 0$$

$$D_B \frac{\partial^2 c_B}{\partial z^2} - b k_2 c_A c_B = 0$$

5.5 KINETICS OF PHYSICAL AND CHEMICAL PROCESSES

There is no clear dividing line between physical and chemical phenomena, so all non-biological processes are considered under this heading.

5.5.1 Solution Equilibria

Many chemical reactions that take place in solution are reversible to some extent. Given sufficient time these reactions tend to reach an equilibrium which can be conveniently defined by a law of mass action coefficient. For example, the ionization of carbonic acid at pH < 8.3 may be represented by

$$H_2CO_3 \rightleftharpoons H^+ + HCO_3^-$$

and the equilibrium may be defined by

$$\frac{[H^+][HCO_3^-]}{[H_2CO_3]} = K_{carbonic\ acid} = 10^{-4.3} \qquad \text{at } 20°C$$

where all the concentrations are expressed in moles. It should be noted that the precise expression of mass action equilibrium refers to activities rather than concentrations. For example, in a reaction defined by the equation

$$mA + nB = pC + qD$$

where A, B, C and D are molecules or ions, and m, n, p and q are coefficients used to balance the equation, the mass action equilibrium is given as

$$\frac{[a_C]^p[a_D]^q}{[a_A]^m[a_B]^n} = K$$

where a is the activity in $mg\,l^{-1}$. Activities are related to concentration by $a = vc$. The activity coefficient, v, is approximately equal to unity for most solutions of non-electrolytes and for dilute solutions of electrolytes (i.e. less than $50\,000$–$100\,000$ mg l^{-1}), so that concentrations may be used in place of activities.

For many inorganic and some organic substances the achievement of equilibrium may be almost instantaneous or within a time span of seconds or minutes. But other reactions may proceed more slowly, so that equilibrium may take months or even years.

Some care is therefore required in the application of equilibrium considerations to water pollution situations. This particularly applies to the use of thermodynamic equilibria in soil–water interactions. Fast reactions proceed so quickly that equilibrium can be assumed in virtually all aquatic environments. Some idea of the speed of fast reactions can be obtained from the data in Table 5.1.

Not all reactions are encounter-controlled (and therefore fast). In the Arrhenius equation

$$k = Ae^{-E_a/RT}$$

where k = rate of reaction
$\quad$ A = frequency factor
$\quad$ E_a = activation energy
$\quad$ R = gas constant
$\quad$ T = absolute temperature

as low reaction is due to a low value of the frequency factor or a high value of the activation energy. Slow reactions are controlled by chemical steps which involve new bonding patterns or solvation, desolvation or configurational changes in entropy. These are required to activate the reactant species to an energetic state from which products can be formed.

Reported values for reversible reactions vary widely from fast to slow reactions:

(a) First order $10^{12}\,M^{-1}\,s^{-1}$ to $10^{-10}\,M^{-1}\,s^{-1}$
(b) Second order $10^{10}\,M^{-1}\,s^{-1}$ to $10^{-11}\,M^{-1}\,s^{-1}$

Even when reactions are irreversible they may take place over such a long period that complete conversion may not be achieved within the retention time in the natural

Table 5.1. Reaction coefficients for fast reversible reactions

Reaction	$K(M^{-1}\,s^{-1})$
$H_3O^+ + OH^- \rightarrow 2H_2O$	1.4×10^{11}
$H_3O^+ + SO_4^{2-} \rightarrow H_2O + HSO_4^-$	1.0×10^{11}
$OH^- + NH_4^+ \rightarrow H_2O + NH_3$	3.3×10^{10}
$HCO_3^- + H_3O^+ \rightarrow H_2CO_3 + H_2O$	4.7×10^{10}

environment. Among the slower irreversible reactions are

(a) metal ion oxidation—notably Fe^{2+} and Mn^{2+}
(b) oxidation of sulphides
(c) metal ion polymerization—notably Al

Some examples of the rates of these slow reactions can be obtained from Table 5.2.

In environments like deep groundwater with very long retention times, equilibrium can be assumed and thermodynamic equilibria provide a useful way of determining the concentrations of a wide range of solutes. For example, the toxicity of aluminium is of interest in many aquatic environments and an indication of the concentration of various species of aluminium may be obtained from thermodynamic considerations of the inorganic reactions involved, which are as follows:

(1) $\dfrac{(Al^{3+})}{(H^+)^3} = K_{Al}$ (2) $\dfrac{(Al(OH)^{2+})(H^+)}{(Al^{3+})} = K_{Al_1}$

(3) $\dfrac{(Al(OH)_2^+)(H^+)^2}{(Al^{3+})} = K_{Al_2}$ (4) $\dfrac{(Al(OH)_2^0)(H^+)^3}{(Al^{3+})} = K_{Al_3}$

(5) $\dfrac{(Al(OH)_4)(H^+)^4}{(Al^{3+})} = K_{Al_4}$ (6) $\dfrac{(AlF^{2+})}{(Al^{3+})(F^-)} = K_{Al_5}$

(7) $\dfrac{(AlF_2^+)}{(Al^{3+})(F^-)^2} = K_{Al_6}$ (8) $\dfrac{(AlF_3^0)}{(Al^{3+})(F^-)^3} = K_{Al_7}$

(9) $\dfrac{(AlF_4^-)}{(Al^{3+})(F^-)^4} = K_{Al_8}$ (10) $\dfrac{(AlF_5^{2-})}{(Al^{3+})(F^-)^5} = K_{Al_9}$

(11) $\dfrac{(AlF_6^{3-})}{(Al^{3+})(F^-)^6} = K_{Al_{10}}$ (12) $\dfrac{(Al(SO_4)^+)}{(Al^{3+})(SO_4^{2-})} = K_{Al_{11}}$

(13) $\dfrac{(Al(SO_4)_2^-)}{(Al^{3+})(SO_4^{2-})^2} = K_{Al_{12}}$

Such a model assumes that the concentrations of Al^{3+} in soil water is in instantaneous equilibrium with the solid phase $Al(OH)_3$ through the reaction

$$3H^+ + 3Al(OH)_3 \rightleftharpoons Al^{3+} + 3H_2O$$

Table 5.2. Half times for some slow irreversible reactions

Reactants	Half time (s)
$Fe^{2+} + O_2$ (pH < 4)	$\approx 10^8$
$Fe^{2+} + O_2$ (pH ≈ 7.5)	10–10^3
$Mn^{2+} + O_2$ (pH ≈ 8)	10^{10}
$H_2S + O_2$	10^2–10^3
$P_2O_7 + H_2O \rightarrow 2PO_4^{3+} + 2H^+$	10^6

and that the Al^{3+} ions formed will then come into equilibrium with other complexation species in a manner determined by K_{Al_1} to $K_{Al_{12}}$.

5.5.2 Chemical Kinetics

As pointed out in Section 5.5.1, equilibrium conditions are not always achieved by reversible or irreversible reactions within the retention time of an environment. The speed of many reactions is slow relative to productive or dispersive processes. An understanding of the time dependence of the reaction is therefore often more important than knowledge of the final equilibrium conditions.

The rate of reaction depends not only on the particular substances involved but also on the physical nature of these substances being present in the same physical state. Heterogeneous systems involve reactions between substances in two or more phases and the rate of reaction may be controlled by factors associated with transference between phases (like surface area) as well as by the rate of chemical reaction.

5.5.3 Homogeneous Systems

Assuming a well-mixed homogeneous system, the manner in which the rate of reaction may vary with the concentration of some or all of the reacting substances is denoted by the order of the reaction. For the reaction

$$A + B \rightarrow P \tag{5.35}$$

The equation for the rate of formation of P can be written

$$\frac{d[P]}{dt} = k[A]^v[B]^w \tag{5.36}$$

where k is the rate constant (per time unit) and $[A]$ and $[B]$ are instantaneous concentrations. The rate at which P is formed varies with time because the concentrations of A and B are reduced as the reaction progresses (Figure 5.7). The form of the curve in Figure 5.7 depends on the number of reacting substances limiting the rate of reaction.

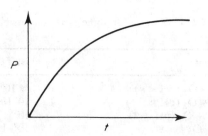

Figure 5.7. Reaction kinetics (the shape of the curve depends on the order of the reaction)

Zero-order Reactions

The reaction rate is independent of the concentration of reacting substances. (More often this occurs in heterogeneous systems due to factors such as surface area available for adsorption limiting the reaction rate.)

If in the reaction $A + B = P$ the exponents v and w in equation (5.36) are zero, then the rate of change of concentration for each reactant may be represented by equation (5.37).

$$-\frac{d[A]}{dt} = -\frac{d[B]}{dt} = \frac{d[P]}{dt} = k \tag{5.37}$$

First-order Reactions

The reaction rate is proportional to the concentration of one of the reactants. For either

$$A \rightarrow B + C$$

or

$$A + B \rightarrow P$$

(where B is present in excess) only [A] matters and the rate of reaction can be represented as

$$-\frac{d[A]}{dt} = k[A] \tag{5.38}$$

which can be integrated to give

$$[A] = [A_0]\, e^{-kt}$$
$$[A_0] = \text{initial concentration at } t = 0 \tag{5.39}$$

[A] is the instantaneous concentration at time t and the amount of [A] consumed in that period is

$$[A_0] - [A] = [A_0](1 - e^{-kt}) \tag{5.40}$$

By substituting $k_1 = 0.4343\, k$ equation (5.39) may be written in the form

$$[A] = [A_0]\, 10^{-k_1 t}$$

or

$$\log_{10} \frac{[A]}{[A_0]} = -k_1 t \tag{5.41}$$

The values of k or k_1 can be found from a plot of $\log [A]/[A_0]$ against t, which should give a straight line of slope k (Figure 5.8) using a least squares analysis. This is a useful method for determining whether a particular reaction is first order or not.

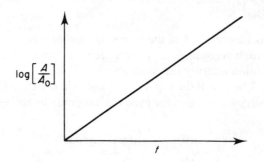

Figure 5.8. Semi-logarithmic plot of a first-order reaction

Values of k can also be determined from the time taken for half-time initial material to react. Substituting the half-life concentration $[A_0]/2$ into equation (5.41) gives

$$\log_{10} \tfrac{1}{2} = -k_1 t_{1/2}$$

$$k_1 = 0.3/t_{1/2} \tag{5.42}$$

Examples of first-order reactions are

(1) the decay of radio-isotopes;
(2) the pseudo-first-order reactions of
   (a) disinfection of organisms in which the number of organisms destroyed per unit or time is proportional to the number of organisms remaining, and
   (b) the hydrolysis of cane sugar in water solution according to

$$C_{12}H_{22}O_{11} + H_2O \rightarrow 2C_6H_{12}O_6 \tag{5.43}$$

   where water is present in such large excess that its concentration does not change significantly during the course of the reaction. The oxidation of organic matter and many other important reactions come into this category.

Second-order Reactions

Two types of second-order reactions can be identified:

(a) $A + B \rightarrow P$

where the concentrations of both A and B affect the reaction rate, which can be represented as

$$-\frac{d[A]}{dt} = k[A][B] = \frac{d[P]}{dt} \tag{5.44}$$

with values of both v and $w = 1$ in equation (5.35).

(b) $A + A = P$

giving

$$-\frac{d[A]}{dt} = k[A]^2 = \frac{dP}{dt} \tag{5.45}$$

$v = 2$ and $w = 0$ in equation (5.36).

For the second-order reaction in which both A and B react, the integrated rate expression takes the form

$$\log_{10} \frac{[B_0]([A_0] - [A])}{[A_0]([B_0] - [B])} = k_1([A_0] - [B_0])t \tag{5.46}$$

The instantaneous concentrations [A] and [B] are equal in a second-order reaction of this type. A plot of the left-hand side of equation (5.46) against t yields a straight line with a slope equal to $k_1 ([A_0] - [B_0])$.

Most reactions of significance in the environment can be approximated by the reaction order described above; however, higher orders and fractional orders are possible. Many proceed in a series of steps, or in competition with other reactions, and many are readily reversible. It is not possible to deduce the order of reaction of the rate limiting step from the stoichiometry of the equation. The order must be determined experimentally.

Complex Reactions

Three classes of complex reactions may be identified:

(1)   Consecutive reactions

$$A \rightarrow B \rightarrow C \text{ (two or more reactions occur in series)}$$

(2)   Back reactions

$$A \rightarrow B \rightleftharpoons P \text{ (readily reversible reaction)}$$

(3)   Competing reactions

$$\left. \begin{array}{l} A + B \rightarrow P \\ A + C \rightarrow Q \end{array} \right\} \text{occurring simultaneously}$$

Complete determination of the rate pattern for complex reactions involves simultaneous solution of the different kinetic equations for each of the individual steps.

For consecutive reactions assuming first-order rate

$$\frac{-d[A]}{dt} = k_1[A] \tag{5.47}$$

$$\frac{-d[B]}{dt} = k_2[B] - k_1 [A] \tag{5.48}$$

$$\frac{d[C]}{dt} = k_2[B] \tag{5.49}$$

Equations (5.47)–(5.49) may be integrated to give

$$[A] = [A_0] e^{-k_1 t} \tag{5.50}$$

$$[B] = [A_0] \left(\frac{k}{k_2 - k_1} e^{-k_1 t} - e^{-k_2 t} \right) + [B_0] e^{-k_2 t} \tag{5.51}$$

$$[C] = [A_0] \left(1 - \frac{k_2 e^{-k_1 t} - k_1 e^{-k_2 t}}{k_2 - k_1} \right) + [B_0] 1 - e^{-k_2 t} + [C_0] \tag{5.52}$$

at time $t = 0$.

If $k_1 \ll k_2$ the reactions will behave kinetically in a manner similar to a single-step reaction.

$$A \xrightarrow{k_1} C$$

When the process for a first-order reaction is readily reversible, such as

$$A \rightleftharpoons B$$

the rate expression includes both the forward and reverse reactions

$$\frac{d[B]}{dt} = \frac{-d[A]}{dt} = k_{+1}[A] - k_{-1}[B] \tag{5.53}$$

at equilibrium

$$\frac{d[B]}{dt} = \frac{d[A]}{dt} = 0$$

and

$$\frac{[A]}{[B]} = \frac{k - 1}{k + 1} = K \qquad \text{where} \quad K = \text{equilibrium constant} \tag{5.54}$$

Comparable relationships can be determined for other complex reactions of different orders.

Temperature Effect

Reaction rates generally increase with increase in temperature. Some approximately double for each 10°C increase.

The Van't Hoff–Arrhenius equation can be used to predict the effect of temperature on the rate coefficient

$$\frac{d(\log k)}{dt} = \frac{E}{RT^2} \tag{5.55}$$

where

E = activation energy (calories) a constant characteristic of the reaction,
R = gas constant (cal per °C),
T = absolute temperature.

Integrating (5.55) between the limits T_0 and T gives

$$\log_e \frac{k}{k_0} = \frac{E(T - T_0)}{RTT_0} \tag{5.56}$$

where k, k_0 are rate constants at T and T_0 respectively.

In aquatic systems with a small temperature change TT_0 does not change significantly and E/RTT_0 can be assumed constant. Equation (5.56) can be approximated by

$$k = k_0 \, e^{C_k(T - T_0)} \tag{5.57}$$

where C_k = temperature characteristic. Often the empirical form is used:

$$k = k_0 \theta^{(T - T_0)} \tag{5.58}$$

where θ = temperature coefficient. Taking the first two terms of the expanded form of e^x, equation (5.57) can be approximated to

$$k = k_0(1 + C_k(T - T_0)) \tag{5.59}$$

5.5.4 Heterogeneous Systems

In heterogeneous systems, the rate of chemical reaction as described in the homogeneous system is further complicated by physical processes resulting from phase boundaries or incomplete mixing. At low concentrations the system may be concentration dependent, but at higher concentrations phase discontinuity may control the reactions.

Three phenomena can be identified: (a) osmosis, (b) diffusion, and (c) adsorption.

(a) *Osmosis* can be defined as the passage of solvent through a membrane from a dilute solution into a more concentrated one. The rate of movement of molecules across the membrane is a function of the concentration at each side. A net movement takes place until either the two concentrations are equal or equilibrium is established by external pressure.

(b) *Diffusion* is the non-advective migration of a substance in solution or suspension in response to a concentration gradient of that substance through another substance. It is a basic natural process which accounts for most of the transport that takes place at the molecular level.

The mechanism is described by Fick's first law, which states that the rate of mass transport by diffusion across an element of unit area is proportional to the concentration gradient of the diffusing substance:

$$N_x = - D_m \frac{\partial c}{\partial x} \tag{5.60}$$

where

N_x = rate of mass transport across an element of area normal to x.

$\dfrac{\partial C}{\partial x}$ = the concentration gradient of the diffusing phase in the x direction.

D_m = the molecular diffusion coefficient, which is proportional to the absolute temperature and inversely proportional to the molecular weight of the diffusing phase and the viscosity of the dispersing phase.

Fick's second law describes the rate of change of concentration of the diffusing phase in an element of the dispersing phase:

$$\frac{\partial c}{\partial t} = D_m \frac{\partial^2 c}{\partial x^2} \tag{5.61}$$

or in three dimensions

$$\frac{\partial c}{\partial t} = D_m \left(\frac{\partial^2 c}{\partial x^2} + \frac{\partial^2 c}{\partial y^2} + \frac{\partial c}{\partial z^2} \right)$$

(c) *Adsorption* occurs when molecules in solution strike the surface of a solid adsorbent and become attached to the surface. At low concentrations of the adsorbate the system may be concentration dependent. Under these conditions the rate of adsorption may be derived as follows:

$$\frac{dc}{dt} = k\phi(c) \tag{5.62}$$

where
  c is the concentration of substance in solution
  t is the time of exposure
  $\phi(c)$ is a function of the concentration in solution
  k is the rate coefficient.

Assuming that $\phi(c)$ can be expressed as $C_0 - C$, the concentration adsorbed, and that k may be modified to allow for the drop in adsorption response during the course of exposure, giving

$$k = \frac{k_0}{1 + r^t}$$

k_0 = rate at $t = 0$
r = coefficient of retardation

then on integrating equation (5.62) and substituting for $\phi(c)$ and k

$$\frac{C}{C_0} = [1 - (1 + rt)^{-k_0/r}] \tag{5.63}$$

At higher concentrations of substance in solution the rate of deposition of molecules on a uniform solid surface may best be described by the Langmuir isotherm, which assumes the formation of a monomolecular layer.

Let x be the fraction of the total solid surface occupied by molecules, $(1 - x)$ being left free at any moment in time. The rate at which molecules are adsorbed is proportional to the available surface area:

$$\frac{dN_a}{dt} = k_1 C(1 - x)$$

and the rate of desorption is proportional to the surface covered:

$$\frac{dN_d}{dt} = k_2 X$$

where k_1 and k_2 are rates of adsorption and desorption and N_a and N_d are the numbers of molecules adsorbed and desorbed. The dynamic equilibrium between free and adsorbed molecules can be expressed by

$$k_1 C(1 - X) = k_2 X \tag{5.64}$$

Rearranging equation (5.63) gives

$$X = \frac{k_1 C}{k_2 + k_1 C} \tag{5.65}$$

If the adsorbed molecules undergo a chemical transformation at a rate proportional to this surface density, then

$$\text{the rate of transformation} = kX = \frac{kbC}{1 + bC}$$

where

$$b = \frac{k_1}{k_2}$$

For small values of C the amount adsorbed is linearly proportional to C while for large values of C the amount adsorbed is independent of C.

5.5.5 Enzyme Reactions

The chemical reactions by which substances are synthesized into cellular materials are good examples of complex reactions taking place in a heterogeneous environment. Transformation of organic molecules in the synthesis of cellular materials depends on the rate of diffusion of material to the cell wall, the rate of adsorption at the cell wall and the various rates of reaction at each step of the complex change within the cell wall. Intermediate compounds are formed which have no cellular functions other than that of opening a pathway to the final cellular molecule, and in a similar fashion many enzyme reactions, take place before a complex organic substance is simplified and returned to more stable organic levels.

The resolution of these complex reactions into simple stages helps to determine the overall rate limiting step which effectively controls the cellular growth. Three advantages of resolution into stages are worth emphasizing.

(1) Simple individual steps may be independently influenced by conditions such as temperature and concentration, leading to a considerable diversity of products in various proportions, giving an appearance of great complexity. Resolution into stages reduces the apparent natural complexity to the level inherent in the system.

(2) The existence of transitory intermediates renders possible the coupling of reactions which appear to be independent from a stoichiometric viewpoint, e.g.

$$B + A \rightarrow BA_2 \qquad \text{with reduction in free energy}$$
$$X + Y \rightarrow X + Y \qquad \text{with increase in free energy}$$

If the reduction of free energy in the former exceeds the increase in the latter then the reactions may be coupled as follows:

$$A_2 \rightarrow A + A$$
$$A + XY \rightarrow AX + Y$$
$$AX + B \rightarrow AB + X$$
$$AB + A \rightarrow BA_2$$

Individual steps may be spatially separated and at a given moment in time not only will concentrations of intermediates exist but also concentration gradients will occur. The overall course of the change depends on the space–time relationship in a variety of ways.

5.6 OTHER PROCESSES

Chemicals, particularly organic substances, may undergo a variety of other processes when discharged into an aquatic environment. It is not possible to give a complete list of all possible processes but the notes below give some examples of approaches that have been adopted for some common problems.

5.6.1 Volatilization

Many organic compounds, like solvents, have a high vapour pressure and may rapidly volatilize through a liquid film. The process may be represented by

$$\frac{1}{K_V} = \frac{1}{K_L} + \frac{1}{HK_G} \tag{5.66}$$

where K_V = volatilization coefficient
$\quad K_L$ = liquid film coefficient
$\quad H$ = Henry's constant
$\quad K_G$ = gas coefficient.

The value of K_L may be calculated from

$$K_L = K_2 \left(\frac{D_L}{D_{O_2}}\right)^{0.5}$$

where k_2 is the reaeration coefficient and D_L and D_{O_2} are the respective diffusivities of chemical and oxygen in water.

The gas coefficient K_G may be calculated from

$$K_G = \frac{0.001}{D} \left(\frac{D_G}{V_G}\right)^{2/3} W \qquad (5.67)$$

where D = stream depth in m
D_G = diffusivity of chemical in air ($cm^2\ s^{-1}$)
V_G = viscosity of air ($cm^2\ s^{-1}$)
W = wind speed ($m\ s^{-1}$)

5.6.2 Hydrolysis

Hydrolysis of chemicals is sometimes pH dependent and many chemicals hydrolyse at significant rates only in acid or alkaline environments. But there are chemicals whose hydrolysis is independent of pH.

The overall rate of hydrolysis may therefore be represented as the summation of the three possibilities.

$$\frac{dC}{dt} = -C(K_{ac}[H^+] + K_{alk}[OH^-] + K_N) \qquad (5.68)$$

where K_{ac}, K_{alk} and K_N are the rate coefficients for acid, alkaline and neutral hydrolysis.

5.6.3 Photolysis

Many chemicals undergo light-induced decay in water when irradiated at wavelengths around 290 nm. The process may be represented by

$$\frac{dC}{dt} = -K_p \theta C \qquad (5.69)$$

where K_p = photolysis rate coefficient
θ = quantum yield.

The photolysis coefficient is dependent on light intensity, depth and turbidity. The effects of turbidity can be further broken down into factors such as colour and suspended solids.

5.7 GROWTH KINETICS

The kinetics of growth are rather more complex than those of chemical reactions, especially when dealing with organisms like invertebrates and fish. Fortunately, the growth of the most important group of micro-organisms—the bacteria—can be represented quite simply and those kinetics with suitable modification may also be used to describe the growth of algae.

For bacteria, the fundamental relationship is between the growth rate and the concentration of substrate. This is shown in Figure 5.9 and may be expressed mathematically as

$$\mu = \mu_{max}\left(\frac{S}{S + K_s}\right) \tag{5.70}$$

where

μ = growth rate
S = substrate concentration
K_s = half-rate concentration as defined in Figure 5.9

The other assumption is that there is a simple relationship between growth and utilization of the substrate which is defined by the yield Y:

$$Y = \frac{\text{weight of bacteria formed}}{\text{weight of substrate utilized}}$$

$$\frac{dX}{dt} = -Y\frac{ds}{dt}$$

These two simplifying assumptions are not completely valid. In particular, the assumption of a constant yield is valid only for a particular bacterial retention time. As bacterial retention time increases, a greater proportion of the substrate is converted to inorganic substances and energy as shown in Figure 5.10.

For algae, the same relationship between growth rate and nutrient concentration has been used by the connection with an inorganic nutrient such as nitrogen,

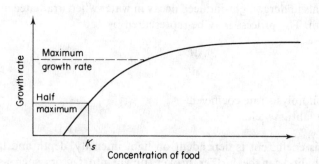

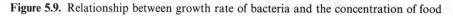

Figure 5.9. Relationship between growth rate of bacteria and the concentration of food

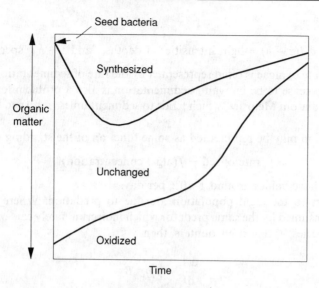

Figure 5.10. Changes to the form of organic matter during a batch test

phosphorus or silicon. The more fundamental process controlling growth is the balance between photosynthesis and respiration. The latter is easily described as a function of biomass

$$\frac{dR}{dt} = RA \qquad (5.71)$$

where

R = rate of respiration
A = algal concentration.

but the rate of photosynthesis is also dependent on light intensity, which varies with time and depth. The relationship with light is generally expressed in terms of the maximum rate of photosynthesis, P_{max}:

$$\frac{dP}{dt} = P_{max}\frac{I}{I_{opt}}\exp\left(1 - \frac{I}{I_{opt}}\right) \qquad (5.72)$$

where

I = light intensity
I_{opt} = light intensity corresponding to P_{max}

Light intensity decreases with depth in an exponential manner at a rate which depends upon the turbidity, thus giving

$$I(z + h) = I(z)\exp(-Kh)$$

where

$I(z)$ and $I(z + h)$ = light intensities at depths z and $z + h$ respectively

The death rate of algae is often represented by the rate of sedimentation, since when algae die, they cease to be buoyant. Sedimentation is also a problem for some living algae, e.g. the diatom Melosira, which tends to sediment unless returned to the surface by advection.

Sedimentation may be represented as some function of the standing crop

rate of sed = f(algal concentration)

where f may have values around 1–10% per day.

Other losses to the algal population are due to predation. Where several prey species are consumed by the same predator which indiscriminately eats only a fraction of the prey species A_i that it encounters, then

$$\frac{dA_i}{dt} = \mu_i g A_i B \tag{5.73}$$

where

μ_i = fraction of species i
A_i = biomass of prey species i
B = biomass of predator.

The growth kinetics of other groups of organisms such as invertebrates and fish may also need to be represented. This is usually done by dividing the population into a number of cohorts (or age groups) on the basis of their fecundity and chances of survival. A matrix of these values may then be used to multiply a vector of numbers in each cohort to determine the future population size and age structure:

$$
\begin{matrix}
\text{matrix of fecundity} & \text{current population} & & \text{new population} \\
\text{and survival} & \text{vector} & = & \text{vector}
\end{matrix}
$$

$$
\begin{bmatrix} f_1 & f_2 & f_3 \\ p_0 & 0 & 0 \\ 0 & p_2 & 0 \\ & A & \end{bmatrix}
\begin{bmatrix} v_1 \\ v_2 \\ v_3 \\ V_1 \end{bmatrix}
=
\begin{bmatrix} v_1 \\ v_2 \\ v_3 \\ V_2 \end{bmatrix}
\tag{5.74}
$$

The values of the fecundities f_1, f_2 and f_3 and the probabilities p_1 and p_2 are density dependent so that the population needs to be incorporated into their calculation, e.g.

$$
A = \begin{bmatrix} 0 & 0.8F & 0.5F \\ 0.35 & 0 & 0 \\ 0 & 0.4S & 0 \end{bmatrix}
$$

where

$$F = \frac{a}{b + N_t} \quad \text{and} \quad S = \frac{1}{1 + e^{N_t/c}}$$

where

$$N_t = \text{population density at time } t$$

a, b, and c are constants.

Reactor Theory

In the treatment of waters and wastewaters unit operations like sedimentation are carried out in specially designed reactors. A similar concept can be applied to the water quality operations that take place in natural environments like lakes.

Reactor theory can therefore be readily applied to modelling the quality changes that occur in rivers, estuaries, lakes, etc. The basic concept is the idea of a mass balance on organisms or material to give

$$\begin{array}{c} \text{mass balance} \\ \text{on X} \end{array} = \begin{array}{c} \text{flux in} \\ \text{of X} \end{array} - \begin{array}{c} \text{flux out} \\ \text{of X} \end{array} \pm \begin{array}{c} \text{rate of change of X} \\ \text{in the reactor} \end{array}$$

The elements of the mass balance—input and output fluxes and rates of reaction—are all closely connected with the hydrodynamic patterns in the reactor. The following notes therefore begin with a consideration of mixing patterns in a reactor.

5.7.1 Types of Reactor

The relationship between influent and effluent quality is determined by the dispersion characteristics of the reactor and the kinetics of the reaction. The two basic types of dispersion characteristics are as follows:

(a) plug-flow, where the liquid flows through the reactor without any mixing taking place;

(b) completely mixed, where liquid entering the reactor is instantaneously mixed with the entire contents.

These may be illustrated by the behaviour of a conservation tracer injected as a pulse into the influent of a reactor. The concentration of tracer in the effluent has two typical shapes as shown in Figure 5.11.

In practice, reactors are encountered that are in varying degrees intermediate between the two kinds. The dispersion characteristics of these intermediate types are most easily represented as a series of completely mixed reactors. The number of reactors in series can be obtained from the following equation:

$$E = \frac{C}{C_0} = \frac{j^j \phi^{(j-1)}}{(j-1)!} \exp(-j\phi) \tag{5.75}$$

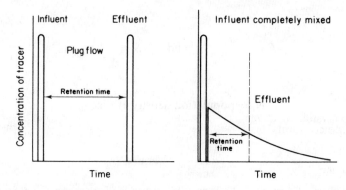

Figure 5.11. Exit-age distribution of a conservative tracer in plug flow and completely mixed reactors

with mean

$$\phi c = 1$$

and variance

$$\sigma_0^2 = \frac{1}{j}$$

$$\bar{t} = \frac{\sum t_c}{\sum C}$$

$$\sigma_t^2 = \frac{\sum t_c^2}{\sum C} - \frac{\sum t_c^2}{\sum C}$$

$$\sigma_0^2 = \frac{\sigma_t^2}{(\bar{t})^2}$$

where

C = effluent concentration
C_0 = initial effluent concentration
$\phi = \dfrac{t}{\bar{t}}$
j = number of equally sized tanks
t = time.

Alternatively, departure from ideality may be expressed by means of the dispersion number. This is based upon the idea of representing a reactor with some degree of dispersion by an equivalent ideal plug flow reactor with a correction factor, i.e.

$$V_{\text{actual}} = V_{\text{plug flow}} \,(\text{correction factor})$$

where the correction factor is some function of

    (i)   intensity of mixing;
   (ii)   reactor geometry;
  (iii)   reaction rate.

The intensity of mixing is generally expressed by the dimensionless group D/uL, where

    D = dispersion coefficient
    u = velocity
    L = length of the reactor.

This is the reciprocal of the Peclet number.

5.7.2 Plug-flow Reactors

Plug-flow reactors may be visualized as containing a series of isolated volumes of liquid which are contiguous with their adjacent volumes but do not mix with them, rather like a series of carriages in a train (Figure 5.12).

The change of concentration during the retention time in the reactor is therefore dependent solely on processes within the individual plugs, and can therefore be represented by a batch reactor. For example, the bacterial and substrate kinetics in a batch reactor are shown in Figure 5.13.

Mathematically, these changes may be given by equation (5.71):

$$\frac{dX}{dt} = \mu_{max}\left(\frac{S}{S + K_S}\right)X - K_d X \qquad (5.76)$$

where

    X = bacterial concentration
    μ = growth rate
    S = concentration of substrate
    K_d = death rate of bacteria

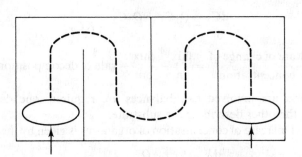

Figure 5.12. Movement of material through a plug-flow reactor

and the substrate kinetics by equation (5.72):

$$\frac{dS}{dt} = \frac{1}{Y} \mu_{max} \frac{S}{S + K_S} X$$

(5.77)

where

$$Y = \text{yield}$$

Plug flow reactors which have only chemical changes may also be described very simply by the appropriate zero-order, first-order or second-order equation.

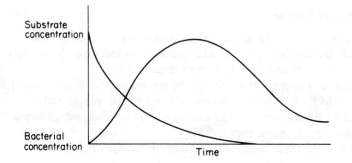

Figure 5.13. Batch system kinetics

5.7.3 Completely Mixed Reactors

Completely mixed reactors are slightly more complicated because bacteria and substrate are continuously being added and removed. They are best described by a mass balance approach. This can be illustrated by considering a completely mixed reactor in which a constituent is undergoing first-order decomposition (Figure 5.14).

The mass balance model can be written as

$$\frac{dC_2}{dt} = \frac{Q_1 C_1}{V} - \frac{Q_2 C_2}{V} - K C_2$$

(5.78)

$$\begin{array}{c}\text{Rate of change of} \\ \text{concentration}\end{array} = \begin{array}{c}\text{flux} \\ \text{in}\end{array} - \begin{array}{c}\text{flux} \\ \text{out}\end{array} - \text{rate of decomposition}$$

When bacteria are involved two balances are required, one dealing with the organisms and the other describing the substrate.

The net rate of increase of concentration of organisms is given by the simple balance

$$\frac{dX}{dt} = \mu X - \frac{Q}{V} X - K \, dX$$

(5.79)

Hence

$$\frac{dX}{dt} > 0 \quad \text{for} \quad \mu > \left(\frac{Q}{V} + K_d\right), \qquad \text{i.e. increase}$$

$$\frac{dX}{dt} = 0 \quad \text{for} \quad \mu = \left(\frac{Q}{V} + K_d\right), \qquad \text{i.e. steady state}$$

$$\frac{dX}{dt} < 0 \quad \text{for} \quad \mu < \left(\frac{Q}{V} + K_d\right), \qquad \text{i.e. decrease}$$

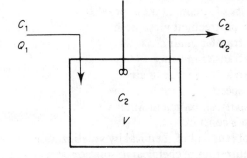

Figure 5.14. Mass balance model of completely mixed reactor with reaction

APPENDIX A

The various processes that have been described in Chapter 5 may be combined in appropriate selections to represent the sinks (and possibly sources) for pollutants in water. The technique of combining these processes with advective and dispersive fluxes is described in Chapters 7–11. The purpose of the appendix is to provide some examples of converting the theoretical representation of a process into a piece of code that can be incorporated into a water quality model. The following three subroutines have been constructed in a modular fashion for two reasons:

(a)   to make formulation and testing more straightforward;
(b)   to make it easy to adapt the model to other parameters by replacing the sinks.

1   Subroutine Volatilization

```
DL = 0.00022*WT**( − 0.67)
DG = 1.9*WT**( − 0.67)
AK2 = (0.00025*(U(I)/1031.0)**0.67)/(D(I)/0.317)**1.85
AKGAS = (0.001/D(I))*((DG/VG)**0.67)*W
AKLIQ = AK2*(DL/DOX)**0.5
AKV = (AKLIQ*AH*AKGAS)/(AKLIQ + (AKGAS*AH))
```

```
      R1(I) = AKV*3600*OCONC(I)*DT
      RETURN
      END


C
C

WT            = atomic mass
DL            = diffusivity of chemical in water
DG            = diffusivity of chemical in air
AK2           = reaeration coefficient
U(I)          = velocity of stream at mesh point I
D(I)          = depth of stream at mesh point I
AKGAS         = mass transfer in air
AKLIQ         = mass transfer in water
VG            = diffusivity of oxygen in air
W             = wind speed
DOX           = diffusivity of oxygen in water
H             = Henry's constant
R1(I)         = rate of removal of chemical by.volatilization
OCONC(I)      = concentration of chemical in solution at mesh point I
DT            = time step
```

Note that numerical values of the variables will have to be transferred into the subroutine through COMMON or the ARGUMENT list. Similarly the value of R1(I) will have to be transferred back to the main program.

2 Subroutine Photolysis

```
C
            IF(T.LT.TS + DY) GOTO 10
            TS = TS + 24.0
   10       IF(T.LE.TS) GOTO 20
            IF(T.GE.TS + DY) GOTO 20
            ETOT = 0.04 + (42.42*AP) + (2.14*ADC) + (0.34*SCONC(I))
            AKPP = AKS*(1.0 − EXP( − 1.25*ETOT*D(I)))
            AKPP = AKPP/(1.25*ETOT*D(I))
            AKPP = AKPP*(1.0 − 0.56*CCOV)
            R3(I) = AKPP*SIN(3.142*(T − TS)/DY)*OCONC(I)*DT
            RETURN
   20       R3(I) = 0.0
            RETURN
            END
```

T   = time
TS   = time of sunrise
DY   = day length
ETOT   = total extinction for the water body
AP   = concentration of plant pigments
ADC   = dissolved organic carbon
SCONC(I)   = suspended solids concentration at mesh point I
AKS   = surface light intensity
AKPP   = underwater light intensity
CCOV   = cloud cover
R3(I)   = rate of removal by photolysis at mesh point I

Note that the subroutine produces a zero value before dawn and after sunset. Between these times the value of R3 varies in a semi-sinusoidal manner.

3   Subroutine Absorption

```
          P6(I) = 0.0
          IF(SCONC(I).LE.O.O) GOTO 202
          DIF = SUL − (OCONC(I) + ADS(I))
          IF(DIF.LE.O.O) GOTO 202
          XCONC(I) = SCONC(I)/1000000.0
          AKPE1 = AKP*(1.0 − 0.5*(EXP( − T/(0.03*AKP))))
          ADSP1 = (AKPE1*ZCONC(I))/(1.0 + AKPE1*XCONC(I))
                  *(OCONC(I) + ADS(I))
          R6(I) = ADSP1 − ADS(I)
          IF(ADSP1.LT.ADS(I)) P6(I) = ABS(R6(I))
          GOTO 203
          IF(OCONC(I).LE.O.O) GOTO 202
C
          GOTO
202       R6(I) = 0.0
203       RETURN
          END
C
```

SUL   = solubility
ADS(I)   = concentration of adsorbed chemical at mesh point I
AKP   = partition coefficient between dissolved and adsorbed species
R6(I)   = transfer of chemical from dissolved to adsorbed form
P6(I)   = transfer of chemical from adsorbed to dissolved form

Note that the subroutine is using equilibrium coefficients to model a dynamic situation. By calculating the potential adsorption at each time step and comparing this with the concentration already adsorbed, it is possible to calculate the amount transferred during each time step.

These subroutines would be called from a major subroutine SINK which could then pass information on rate of uptake back to the main program.

It is extremely important that all variables and subroutines are consistent in their units.

Chapter 6

Kinetic Modelling and Statistics

P. M. BERTHOUEX

6.1 INTRODUCTION

Kinetic modelling involves stating a model which is to be tentatively assumed adequate. The statistical problems are, first, to estimate the parameters in the model (the dispersion coefficient, rate coefficients, mass transfer coefficients, etc.) and then to check the adequacy of the model, i.e. to test the assumption that the assumed structure of the model was satisfactory in light of the available data. If the model is inadequate in any respect, it needs to be modified and refitted to the data. Perhaps additional data will be needed. The overall process of model building is iterative—collect data, estimate parameters, check model adequacy, modify the model or the experimental design, and do more experiments.

The model builder wants to discover the true model. Statistics can, at best, only make inferences about the truth. Measurements are imperfect. Not all influential variables are known, let alone measured. These difficulties, and others, prevent us from ever learning the true model exactly. By making carefully planned measurements and using them properly, knowledge is gradually increased until a model is found that would satisfy statistical and practical considerations.

Statistics helps us move toward the truth, but it cannot guarantee that we will reach it, nor will it tell us whether we have done so. It can help us make honest scientific statements about the likelihood of certain hypotheses being true.

6.2 THE LEARNING PROCESS

Learning is an iterative process, the key elements of which are shown in Figure 6.1. The cycle begins with expression of a working hypothesis, which is typically based on

An Introduction to Water Quality Modelling. 2nd Edition. Edited by A. James
© 1993 John Wiley & Sons Ltd.

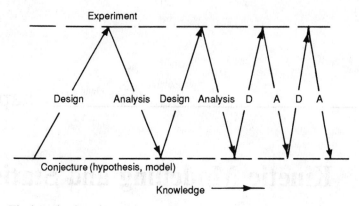

Figure 6.1. The iterative learning cycle

a priori knowledge about the system. The hypothesis is usually stated in the form of a mathematical model. Whatever form the hypothesis takes, it must be probed and given every chance to fail as data become available.

Learning progresses most rapidly when the experimental design is statistically sound. If it is poor, little will be learned. A statistically efficient design literally may let us learn as much with eight experiments as from eighty. Iterating between experimentation (data collection) and hypothesis testing (data analysis) provides the opportunity for improving precision by shifting emphasis to different variables, making repeated observations, and adjusting experimental conditions.

An experiment is like a window through which we view nature (Box, G. E. P., 1974). Our view is never perfect. The observations that we make are distorted by imperfections, which are called 'noise'. This idea is sketched in Figure 6.2. A statistically efficient design reveals the magnitude and characteristics of the noise. It increases the size and improves the clarity of the experimental window. Using a poor design is like seeing blurred shadows behind the window curtains. Analysing happenstance data (data collected with no regard for experimental design) may be like looking out of the wrong window.

As much as we prefer working with experimental designs that are statistically designed, it is not always possible. We may need to assess changes that have occurred over time and the available historical data were not collected with a view toward assessing these changes. In natural systems, treatment plants, and in some industrial processes, we cannot set and manipulate the independent variables to create conditions of special interest. A related problem is not being able to replicate experimental conditions. These are huge obstacles to model building. Hopes for successfully extracting information from such happenstance data are often not fulfilled.

6.3 SPECIAL PROBLEMS

Introductory statistics books commonly deal with linear models and assume that the available data are normally distributed and independent, with constant variance.

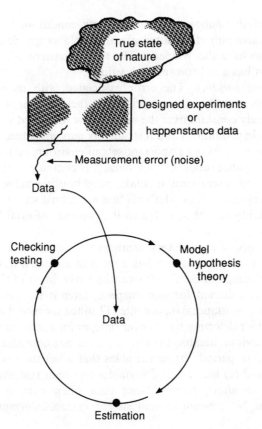

Figure 6.2. An experiment provides a view of nature, represented by data which are imperfect observations because they contain measurement error (noise). The data are subjected to the iterative statistical learning cycle

Often, in water quality modelling, the data do not satisfy these fundamental assumptions, or nonlinear models are needed. Some specific problems encountered in data acquisition and analysis are as follows:

(1) *Large measurement errors.* Counts of bacterial populations have large measurement errors. So do many chemical measurements, despite the usual care that is taken with instrument calibration, reagent preparation and personnel training. A portion of this error derives from sampling non-homogeneous materials.

(2) *Aberrant values.* Values that stand out from the general trend are fairly common. They may reflect gross errors in sampling or measurement, or mistakes in data recording. If we think only in these terms, it becomes too tempting to throw out such values. However, the unusual value may be real, in which case it contains information of great importance.

(3) *Non-normal distributions.* We are strongly conditioned to think of data as being symmetrically distributed about their average. Sets of environmental data seldom have this property. The typical pattern is a positive skew (the distribution has a tail toward high values).

(4) *Non-constant variance.* The error associated with measurements is often nearly proportional to the magnitude of their measured values rather than approximately constant over the range of the measured values. Measurement procedures (e.g. dilution) and instruments may introduce this property.

(5) *Serial correlation.* Many important sets of environmental data occur as time series or as spatial series. In such data, it is common that natural continuity over time and space tend to make neighbouring values more alike than randomly selected values. Methods that are robust with respect to deviations from normality can be sensitive to the presence of even low levels of serial correlation.

(6) *Large amounts of data.* Every treatment plant, river basin authority and environmental control agency has a mass of data that has collected in filing cabinets or computer databases over the years. Most of this is happenstance data. It was collected for one purpose; later it is considered for another purpose. Happenstance data are often ill suited for model building. They may be ill suited for detecting trends over time, or for testing any hypothesis about system behaviour because (a) the record is not consistent and comparable from period to period, (b) all variables that affect the system have not been observed, and (c) the range of variables has been restricted by the system's operation. In short, happenstance data often contain surprisingly little information. No amount of analysis can extract information that does not exist.

(7) *Lurking variables.* Sometimes important variables are unknown or measured, for a variety of reasons. Such variables, called 'lurking variables', can cause serious problems (Box, G. E. P. 1966; Joiner 1981).

In order to gain from what statistics offers we must proceed with an attitude of letting the data reveal its critical properties and of selecting statistical methods that are appropriate to deal with these properties.

6.4 BASIC STATISTICAL CONCEPTS

Repeated observations of some physical, chemical or biological characteristic that has the true value η will not be identical, even though the experimenter has tried to make the experimental conditions as identical as possible. The discrepancy is allowed for in the equation $y_i = \eta + e_i$, where y_i is the observed value and e_i is an error or disturbance. Each quantitative result should be reported with an accompanying estimate of its error.

Error (also called experimental error or noise) in the statistical context does not imply fault, mistake or blunder. It refers to variation that is often unavoidable, resulting from such things as measurement fluctuations due to instrument conditions, sampling imperfections, variations in ambient conditions, skill of personnel and many other factors.

6.4.1 The Average and the Standard Deviation

A statistic is a quantity calculated from a set of data that is taken to represent a population. A parameter is a quantity associated with the population. Parameters cannot be measured directly unless the entire population can be observed; they are estimated by statistics.

A sample of n observations is actually available from the population used to calculate the sample average: $\bar{y} = \sum y/n$. The statistic estimates the mean of the population, η.

The true value of the variance, σ^2, is estimated by the sample variance $s_y^2 = \sum (y - \bar{y})^2/(n - 1)$, where n is the size of the sample and $\bar{y}$ is the sample average. The sample standard deviation is the square root of the sample variance. The denominator, $n - 1$, is the number of degrees of freedom of the sample.

6.4.2 The Normal Distribution

Repeated observations that differ because of experimental error often vary about some central value with a bell-shaped probability distribution that is roughly symmetric and in which small deviations occur much more frequently than large ones. The continuous distribution that represents this is the Gaussian or normal distribution.

There is a tendency for error distributions that result from many additive component errors to be 'normal-like'. This is the central limit effect, which rests on there being several sources of error, where no single source dominates, and the overall error being a linear combination of independently distributed errors. When this holds, the distribution or errors tend toward normality even when the contributing component errors are not normally distributed.

The normal distribution does not describe all data, sometimes it is not even an adequate approximation. Environmental data are often not normally distributed because some component errors are multiplicative and not additive. Dilution, for example, creates a multiplicative effect. A common distribution is the positive skew. Often the logarithms of such data are normally distributed.

Many commonly used statistical procedures are robust or insensitive to deviations from normality. This means that they yield correct conclusions even when the data are not normally distributed. The normal distribution is characterized completely by the mean and variance (or standard deviation). The standard deviation σ measures the distance from the mean to the point of inflection of the curve. The probability that a positive deviation from the mean will exceed one standard deviation (1σ) is roughly 1/6. The probability that a positive deviation will exceed 2σ is roughly 1/40. The chance that a deviation in either direction will exceed 2σ is roughly 1/20 or 0.05.

6.4.3 The Sampling Distribution of the Average

The statistics $s_{\bar{y}}$ and s^2, and all other calculated statistics, have an expected value (mean) and a variance. Here we consider the variance of the average $\bar{y}$.

For the sample, $s_y^2 = \sum(y - \bar{y})^2/(n - 1)$ and $s_{\bar{y}} = s/\sqrt{n}$. The standard deviation of the mean, s_y, often called the standard error to distinguish it from the sample standard deviation.

If the parent distribution is normal, the sampling distribution of $\bar{y}$ will be normal. If the parent distribution is non-normal, the distribution of $\bar{y}$ will be more nearly normal than the parent. As the number of observations n used in the average increases, the distribution of y becomes increasingly more normal. This is due to the central limit effect.

6.4.4 The Distribution of s^2 and t

If the parent distribution is normally distributed and has mean η and variance σ^2, then: (1) the sample variances s^2 is distributed independently of $\bar{y}$ as a scaled χ^2 (chi-square) distribution (the scaled quantity is $\chi^2 = vs^2/\sigma^2$) and (2) the quantity $t = (y - \eta)/(s/\sqrt{n})$ has a t distribution with $v = n - 1$ degrees of freedom, which is a symmetric distribution and the spread decreases as the number of degrees of freedom of s increases. If the parent population is not normal, but the sampling is random, the distribution of the t statistic tends toward the t distribution, just as the distribution of y tends toward being normal.

6.4.5 Confidence Intervals

It is informative to state an interval within which the value of a parameter would be expected to lie. A $1 - \alpha$ confidence interval can be constructed, using the appropriate value of t, as $\bar{y} - s_{\bar{y}}t_{\alpha/2} < \eta < \bar{y} + s_{\bar{y}}t_{\alpha/2}$, where $t_{\alpha/2}$ and $\bar{y}$ have $v = n - 1$ degrees of freedom.

The meaning of the $1 - \alpha$ probability is 'If a series of random sets of n observations is sampled from a normal distribution with mean η and fixed σ, and $1 - \alpha$ confidence interval $y \pm t\alpha/2s_y$ is constructed from each set, a proportion $1 - \alpha$ of these intervals will include the value η and a proportion α will not (Box *et al.* 1977).' The Bayesian interpretation is that there is a $1 - \alpha$ probability that the true values fall within this confidence interval.

6.5 PARAMETER ESTIMATION—THE METHOD OF LEAST SQUARES

A linear model, $\eta = f(x, \beta)$, or nonlinear model, $\eta = f(x, \theta)$, with independent variables x, is to be fitted to observations y_i in order to estimate a vector of parameters (β or θ). The true underlying value, η, and the observed value, y, differ because of measurement error. They are related as follows: $y_i = \eta + e_i$, where e_i is an independent, normally distributed random error having constant variance.

Linear and non-linear, in this context, refer to the parameters and not to x, because in parameter estimation x is known and β is not. A model is 'linear in the parameters' if the equation obtained by taking derivatives with respect to all parameters is linear in the parameters. For example, the curvilinear model $\eta = \beta x^2$ is said to be linear because $\partial\eta/\partial\beta = 2x$ is linear in the parameter β. That it is nonlinear with respect to x is irrelevant because a numerical value exists for x after the experiment has been done. In contrast, $\eta = \exp(-\theta x)$ is nonlinear in θ because $\partial\eta/\partial\theta = -\theta \exp(-\theta x)$ is nonlinear in θ.

Regardless of the form of the model, the best (unbiased and minimum variance) estimates of the parameters are obtained using the method of least squares:

$$\min S = \sum(e_i)^2 = \sum(y_i - \eta_i)^2$$

For unbiased estimates, the parameter estimates converge towards their true values as the number of observations increases.

For linear models, a simple algebraic solution exists. For example, to estimate β in $\eta = \beta x^2$, we need to find

$$\min S = \sum(y_i - \beta x_i^2)^2 = \sum(y_i^2 - 2\beta x_i^2 y_i + \beta^2 x_i^4)$$

The least squares estimate of β makes $\partial S/\partial\beta = 0$. The reader can confirm that the solution is $\beta = \sum x_i^2 y_i / \sum x_i^4$. Computer programs use linear algebra to solve these linear least squares (regression) problems. The key point is not how the computations are done, but that a unique algebraic solution exists for linear models.

No unique algebraic solution exists for nonlinear models. Obviously for

$$\min S = \sum[y_i - \exp(-\theta x_i)]^2 = \sum[y_i^2 - 2y_i \exp(-\theta x_i) + \exp(-2\theta x_i)]$$

the derivative $\partial S/\partial\theta$ will contain an exponential function of θ and the value of θ that minimizes S must be found by numerical search.

6.6 JUDGING THE ADEQUACY OF A MODEL

The first step is to plot the prediction of the model and the data against the independent variable. Next plot the residuals ($e_i = y_i - \eta_i$) against the predicted values and against each of the independent variables. If the model fits, the residuals will be random variables and no patterns or trends will exist. Patterns are evidence of a weakness in the model. In addition to serving this diagnostic purpose, residual plots also give values as to how the inadequacy might be removed (Draper and Smith 1981).

The minimum value of SS found by linear or non-linear least squares is the residual sum of squares, RSS. If the model is correct and truly fits the data the RSS consists only of random measurement errors and an estimate of the 'pure measurement error' variance, $s^2 = \text{RSS}/(n - p)$. If the model does not fit, the RSS is inflated because then it consists of pure error plus error due to lack of fit.

If the experiment includes repeated measurements at some settings the pure error can be estimated independently and a lack of fit test can be made. Without replication, lack of fit is judged by plotting the fitted model against the data and examining the residual errors for signs of model inadequacy.

Assume that the model fits and the RSS reflects only measurement error. If the experiment were repeated, a new series of observations would contain different errors with the result that the minimum sum of squares, RSS, and the parameter estimates would change from experiment to experiment. For a given set of data, SS could be calculated over a grid of parameter values to generate a surface. Contour maps of its projection onto a plane also could be plotted. Such maps are commonly used to display the joint confidence region of the parameters.

6.7 JOINT CONFIDENCE REGION OF THE ESTIMATED PARAMETERS

Calculating the 'best' values of the parameters is only half the job of fitting and evaluating a model. The precision of these estimates must be evaluated. This aspect of the job too often is overlooked, perhaps because computing the joint confidence region is considered difficult. Fortunately, it is not.

The precision of estimated parameters, in a linear or nonlinear model, is indicated by the size of the joint confidence region. 'Joint' indicates that the parameters in the model are considered simultaneously.

A small joint confidence region indicates precise parameter estimates. The orientation and shape of the confidence region also are important. In general the size of the confidence region decreases as the number of observations used increases. Precision, however, also depends on the actual choice of settings of independent variables at which measurements are made. In practice, especially with nonlinear models, these settings sometimes are more important than the number of observations.

6.7.1 A Linear Model

Assume that the model $\eta = \beta_0 + \beta_1 x + e$ has been fitted by the method of least squares to give the fitted model $y = b_0 + b_1 x$. The estimates b_0 and b_1 are normally distributed random variables with means equal to β_0 and β_1. The $100(1 - \alpha)\%$ confidence intervals for β_0 and β_1 are

$$\beta_0: \quad b_0 \pm t_{n-2,\alpha/2} s[(1/n) + (x^2/s_x^2)]^{1/2}$$

$$\beta_1: \quad b_1 \pm t_{n-2,\alpha/2} s[(1/S_x^2)]^{1/2}$$

where $s = [\sum(y_i - b_0 - b_1 x_i)^2/(n - 2)]^{1/2}$ and $S_x^2 = \sum(x_i - x)^2/n$. When the probability level α is set at 0.05, these equations define the 95% confidence intervals for the parameters in question.

The joint confidence region for both β_0 and β_1 is not provided by the simultaneous use of the individual confidence intervals (sometimes called the marginal confidence limits). The $100(1 - \alpha)\%$ joint confidence region for β_0 and β_1 is enclosed by an ellipse, such as shown in Figure 6.3. The individual confidence limits, shown by straight lines at the intervals marked on the axes, suggest that the cross-hatched rectangle might represent the precision of the parameters. This is wrong. The shaded ellipse is the joint confidence region (Hunter 1981).

Consider a point in the region marked A. This is within the joint confidence region (the shaded area) but would be rejected by the individual intervals (i.e. inside the shaded ellipse but not inside the cross-hatched rectangle). Note that this is also true for a portion at each end of the joint confidence region.

The opposite problem also exists, as indicated by the region marked B. Here a pair of parameter values is outside the joint confidence region and should be rejected, but would not be when viewed (incorrectly) in terms of the two individual confidence intervals.

The joint confidence region for a linear model will always be an ellipse or, in higher dimensions, an ellipsoid. The least squares parameter estimates will be centrally located in the ellipse. This simple and predictable geometry does not exist for non-linear models.

6.7.2 A Nonlinear Model

Exponential functions are common in environmental modelling. An example is the classical BOD model: $y_i = L(1 - \exp(-kt_i)) + e_i$. The parameters to be estimated are, L, the asymptotic BOD (ultimate BOD) that is approached as time becomes large, and k, the first-order reaction rate coefficient.

Many published papers show estimates of L and k derived from measurements at just a few early times, say days 1 through 5. This, we shall see, gives very poor parameter estimates. Some measurements at 15–20 days are needed to define the asymptotic level.

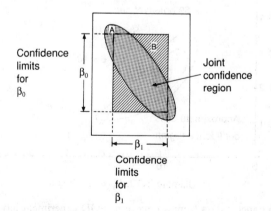

Figure 6.3. Individual confidence intervals for the parameters β_1 and β_2 and their ellipsoidal joint confidence region

The data shown in Figure 6.4, 59 observations on wastewater from a dairy, were fitted using non-linear least squares to estimate $L = 10\,100$ mg/l and $k = 0.217$ day^{-1}. The line was computed using these parameters values. The 95% joint confidence region is plotted, also in Figure 6.4. The interpretation is that there is a 95% probability that the true parameters values are within this region. The small size shows that both parameters are estimated with reasonable precision. The orientation of the region indicates a slight negative correlation between the parameters. (If L is

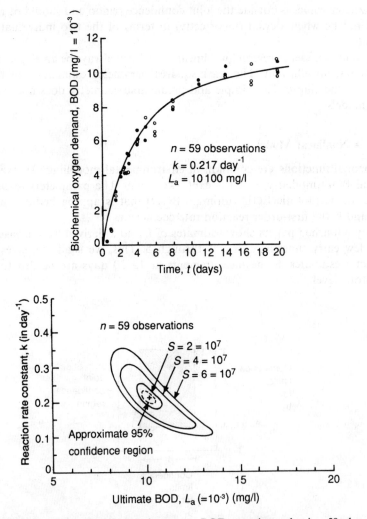

Figure 6.4. The top panel is data from a long-term BOD experiment having 59 observations shown with the fitted model: $y = 10100[1 - \exp(-0.217\ t)]$. The bottom panel shows the approximate 95% joint confidence region for the estimated parameters

increased, the value of k must be increased slightly in order to still fit the data as well as possible.) The parameter correlation is slight and we would be willing to consider that nearly independent estimates of each parameter have been obtained.

Figure 6.5 shows what happens using 30 observations on days 1 to 5 and no measurements on the asymptote. Neither parameter is estimated very well, as shown

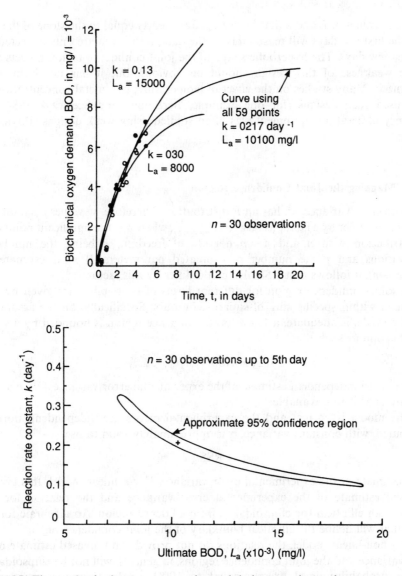

Figure 6.5. The result of using a poor experimental design to estimate the parameters in the BOD model; the joint confidence region is elongated and the parameter estimates are so imprecise as to be virtually useless

by the large, elongated, hyperbolically shaped joint confidence region. Almost any value for L can be used if a corresponding reduction is made in k. Thirty observations, a great deal of work, have been wasted. The problem is that the data only allow estimation of the initial slope of the curve, which is

$$\left.\frac{dy}{dt}\right|_{t=0} = Lk \exp(-k)|_{t=0} = Lk = \text{constant}$$

Thus any values of L and k that have a product nearly equal to the slope of the curve over the first few days will reasonably fit the data, if the data are only collected over the first few days. The hyperbolic shape of the joint confidence region reflects this.

The weakness of this design should be obvious, but it has not been widely recognized. Many studies on the effect of temperature, pH, metal concentration, etc. have used such designs (for one example, Berthouex and Scewczyk 1984). The efficiency of aeration equipment was often studied using weak designs (Boyle *et al.* 1974).

6.7.3 Mapping the Joint Confidence Region

The pure error variance, s^2, has an F distribution. Therefore, s^2 values from different experiments differ by a factor as large as $F_{\alpha(n,n-p)}$, where α is the significant point of the F distribution with n and $n - p$ degrees of freedom, n being the number of observations and p the number of estimated parameters. If s^2 is estimated by $\text{RSS}/(n - p)$, it follows that RSS also could differ by this factor.

The joint confidence region for different levels of probability are given by areas contained within specific sum of squares contours. Specifically, on the assumption that the model is adequate, a $1 - \alpha$ joint confidence region is bounded by a sum of squares contour such that

$$S_{1-\alpha} = \text{RSS} + s^2 p F_{\alpha,(p,n-p)}$$

where s^2 is an independent estimate of the experimental error variance. Often a direct estimate of s^2 is not available.

If the model is correct and the experimental errors are independent, normally distributed, with constant variance, it is approximately valid to use

$$s^2 = \text{RSS}/(n - p)$$

as an estimate of the experimental error variance. For a linear model, this gives an unbiased estimate of the experimental error variance and the relation for $S_{1-\alpha}$ describes an elliptical (or ellipsoidal) joint confidence region. An algebraic formula exists that will define the elliptical boundary of the joint confidence region.

For a non-linear model, the relations do not provide an unbiased estimate or the error variance and the joint confidence regions, in general, will not be ellipsoidal nor have a probability level exactly equal to $100(1 - \alpha)\%$. Common practice, for nonlinear models, is to call the regions just defined 'approximate joint confidence regions' (Draper and Smith 1981).

Plotting the nonlinear least squares joint confidence regions is a simple brute force computing job. The sum of squares values is calculated using many different combinations of parameter values, the $100(1 - \alpha)\%$ sum of squares level is determined, and a contour line is interpolated so that all grid points having a sum of squares lower than this critical level are inside the joint confidence region.

6.7.4 Comments

All reports of estimated parameters should include a statement regarding their precision. The best way to make this statement is to plot the joint confidence region. This gives a quick and accurate impression of the size, shape and orientation of the confidence region, all of which are important. It also prevents wrong conclusions being reached by using the individual (marginal) confidence intervals.

The size, shape and orientation of the joint confidence region are all fixed by the data. They cannot be changed by any valid method of data analysis. The location of observations can be as important as the number of observations. In fact, a few well-placed observations may give parameter estimates that are more precise and less correlated than a massive, but poorly designed, effort.

6.8 WHY LINEARIZATION GIVES POOR PARAMETER ESTIMATES

An experimenter invests care, time and money to obtain data and should want to extract all the information the data contain. If this purpose is to estimate parameters in a nonlinear model, he should insist that the parameter estimation method gives estimates that are (1) unbiased and (2) precise. Generally, the best method of estimating the parameter will be nonlinear squares, in which variables are used in their original form and units.

Unfortunately, some experimenters transform the model so it can be fitted by linear regression. This can, and often does, give biased or imprecise estimates. The dangers of linearization will be shown by examples.

6.8.1 Example—BOD Data

The Thomas slope method has been used by many engineers over the years to linearize the classical first-order BOD model so it can be plotted as a straight line or fitted by linear regression. Figure 6.6 compares parameter estimates obtained from the Thomas slope method with those computed using nonlinear squares (NLS). The approximate 95% joint confidence region, computed from the NLS results, is also plotted (Berthouex and Scewczyk 1984).

This comparison was made for 20 sets of BOD data. In every case, the TSM estimates were biased. In 12 cases, the TSM estimates did not fall within the approximate 95% joint confidence region of the NLS estimates. The Thomas slope

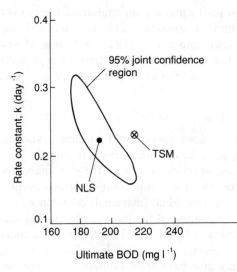

Figure 6.6. The parameter estimates obtained using the method of nonlinear least squares (NLS) and their approximate 95% joint confidence region compared with the badly biased parameter values estimated using the Thomas slope method (TSM), which fall far outside the 95% confidence region

method may have been useful before nonlinear least squares estimates could be done easily. Today it should never be used.

Why is the TSM so bad? The method involves plotting $Y_i = \sqrt[3]{1/y_i}$ on the ordinate against $X_i = y_i/t_i$ on the abscissa. These transformations distort both X and Y. Each X_i contains error in different amounts, depending on the error in y_i. In linear regression, the X variable is supposed to be free of error or to at least have an error that is very small compared to the errors in the Ys. Also, the error structure of the ordinate, $Y_i = \sqrt[3]{1/y_i}$, is badly distorted, first by taking by the reciprocal and then by the cube root, and the distortion gets worse as the range of y values increases. Thus, the condition of constant variance, upon which linear regression rests, is obliterated by the transformation.

A collection of linearization methods has been developed for fitting kinetic data, more than a half dozen for the simple BOD reaction alone. All carry the danger of distortion illustrated with the Thomas method. They should not be used unless it can be shown that the transformations better fulfils the assumptions on which regression is based.

6.8.2 Case Study—Michaelis–Menten Model

The Michaelis–Menten model states, in biochemical terms, that an enzyme-catalysed reaction has a rate $\eta = kX/(K + X)$, where X is the concentration of substrate (the independent variable). The observed values are

$$y_i = \eta + e_i = \frac{kX_i}{K + X_i} + e_i$$

Three ways of estimating the two parameters in the model are:

(1) non-linear least squares, which will give the best parameter estimates. This is found by searching on K and k to

$$\text{minimize } S = \sum_{i=1}^{n} \left[y_i - \frac{kX_i}{K + X_i} \right]^2$$

(2) The Lineweaver–Burk plot, also known as the double reciprocal plot. This method is still widely used, which is unfortunate because, of the three methods, it gives the worst parameter estimates. The model is rearranged to give

$$\frac{1}{y_i} = \frac{K}{kX_i} + \frac{1}{k}$$

A plot of $1/y_i$ against $1/X_i$ should be linear with intercept $1/k$ and slope K/k.

(3) Linearization using y_i against y_i/X_i. This method is better than the Lineweaver–Burk plot, but it gives biased estimates. The transformed model, which has intercept k and slope K, is

$$y_i = k - K\frac{y_i}{X_i}$$

To show the deficiencies of the two linearization methods, Coloquhoun (1971) estimated the parameters in this model by all three methods for 750 simulated experiments. Each simulated experiment used five 'observations' that were created by adding normal random errors (with mean zero and $\sigma^2 = 1$), to values calculated using the model with $k = 30.00$ and $K = 15.00$. Thus, unlike what happens in a real experiment, the distribution, mean, and standard derivations are known and the 'data' satisfy the assumptions on which the method of least squares rests. The resulting 750 estimates of k were grouped to form histograms and are shown in Figure 6.7.

The distributions for K are similar since the estimates of k and K are highly correlated. Experiments that yield an estimate of k that is too high tend to give an estimate of K that is too high also, whichever method of estimation is used. The parameter correlation is a result of the model structure, perhaps emphasized or inflated by the location (spacing) of the X observations.

The NLS estimates are more closely grouped around the true value ($k = 30.00$) than the estimates found by the other methods. The average value of the NLS estimates is 30.4, close to the true value. They have little bias. The NLS have the smallest variance.

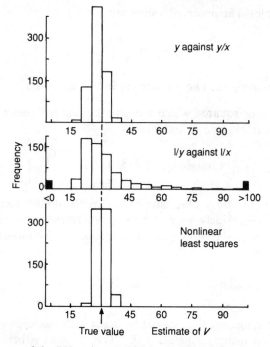

Figure 6.7. Histograms of the 750 estimates of k in the Michaelis–Menten equation obtained using three parameter estimation methods in 750 simulation experiments. The 'true value' of k was 30. The Lineweaver–Burk plot is the middle histogram, which is much worse than the other methods. The bottom histogram shows the estimates obtained using the method of nonlinear least squares; these are the most precise and least biased

By comparison, the Lineweaver–Burk method gives terrible estimates. The scatter is much greater. Near infinite estimates are obtained when the plot goes nearly through the origin, giving $1/k = 0$. These distort the average of the estimated values so much that no realistic estimate of the bias is possible.

The method of plotting y against y/X gives the estimates falling between these extremes. Their standard deviation is only about 28% greater than that of the NLS estimates. About 73% of the estimates were too low (below 30) and the average was 28.0. They have a negative bias. This bias is purely a consequence of the estimation.

The results of Coloquhoun's simulation are convincing about linearization being a risky procedure. The results are not widely applicable. The histograms will look different for different sorts of errors, different distributions for the observations, and different experimental designs (spacing and number of X values). The y against y/X method was better than NLS for the case where the coefficient of variation (standard deviation/mean) was the same at each X value. The Lineweaver–Burk method was still awful (Coloquhoun 1971).

6.8.3 Comments

Linearization will give good parameter estimates when the error structure (variance) of the observations fortuitously matches the particular linearizing transformation.

For example, a log transformation is helpful when var(y_i) is proportional to y_i. These situations do arise. So, we do not say 'never linearize', but we do say 'don't linearize merely to create a linear regression problem'. When it is inappropriate, the consequences can be severe. Learn to use nonlinear least squares. The computations are easily managed with any unconstrained minimization program.

6.9 EXPERIMENTAL DESIGN FOR NONLINEAR PARAMETER ESTIMATION

The immediate problem is to design experiments that will yield precise parameter estimates with a minimum of work and expense. 'Designing the experiment' means specifying how many observations will be made and at what settings of the independent variable they will be made.

6.9.1 The Experimental Design Method

The general nonlinear model is

$$\eta = f(\theta_i, \xi_u), \qquad i = 1, 2, \ldots, p; \quad u = 1, 2, \ldots, n$$

where θ_i are the parameters to be estimated (rate coefficients, etc.) and ξ_u are the independent variables (time, temperature, dose, etc.). The parameters will be estimated by nonlinear least squares. If the model contains p parameters, the minimum number of observations that will yield parameter estimates is $n = p$. The model in general will not pass through these points (unlike a linear model with $n = p$, which will fit each observation exactly).

When the model is linear the variances and covariances of the parameters are exactly

$$\text{var}(\theta) = \{X'X\}^{-1} \sigma^2$$

and approximately this when the model is nonlinear. The size of the joint confidence region is minimized by minimizing var(θ).

Since the variance of the random measure error, σ^2, is a constant (even though its value may be unknown), it is only the $\{X'X\}^{-1}$ matrix that need be considered. Rather than trying to compare the entire variance–covariance matrices for different designs, simply minimize the determinant of the $\{X'X\}^{-1}$ matrix or equivalently maximize the determinant of $\{X'X\}$. This latter design criterion, presented by Box and Lucas (1959), is written as

$$\max \Delta = |X'X|$$

where the vertical bars indicate the determinant. For nonlinear models, this results in minimizing the approximate joint confidence region. The size of the approximate confidence region is inversely proportional to the square root of the determinant; $\Delta^{-1/2}$.

X is an $n \times p$ matrix, called the derivative matrix:

$$X = \{X_{iu}\} = \begin{bmatrix} X_{11} & X_{21} & \cdots & X_{p1} \\ X_{12} & X_{22} & \cdots & X_{p2} \\ \vdots & \vdots & & \vdots \\ X_{1n} & X_{2n} & \cdots & X_{pn} \end{bmatrix} \quad i = 1, 2, \ldots p; \quad u = 1, 2, \ldots n$$

where p and n are the number of parameters and observations as defined earlier. The elements of the X matrix are partial derivatives of the model with respect to the parameters:

$$X_{iu} = \partial f(\theta_i, \xi_u)/\partial \theta_i \quad i = 1, 2, \ldots, p; \quad u = 1, 2, \ldots, n$$

For non-linear models, the elements X_{iu} are functions of θ, the true values of which are unknown; in fact, it is because they are unknown that experiments are needed. This seems to say that in order to design efficient experiments we must know in advance the answer that those experiments will give. The situation is not hopeless, however, because the experimenter, based on his own knowledge and perhaps on previous similar experiments, has some prior knowledge concerning what reasonable parameter values might be. He can provide estimates that hopefully are not too remote from the values. Expecting that this will be the case, these *a priori* estimates θ^* are used to evaluate the elements of the derivative matrix and design the first experiments.

The design is optimal with respect to the 'guessed' parameter values, and based on the critical assumption that the model is correct. Here 'optimal' has its precise mathematical meaning. In reality, the experiment is not likely to be optimal in the sense that its results will completely satisfy the experimenter. If the initial parameter 'guess' is not close to the true underlying value, the confidence region will be large and more experiments will be needed. If the model is incorrect, the experiments planned using this criterion will not reveal it.

The so-called 'optimal' design, then, should be considered as advice that should make experimentation more economical and rewarding. It is not a prescription for getting perfect results with the first set of experiments. Because of this, an iterative approach to experimentation is often planned.

The iterative experimental procedure is extremely efficient. Even if the initial design gives relatively poor parameter estimates, it will guide the experimenter quickly to more informative investigations. Usually, in these iterations, the joint confidence region is very small. Of course, data from all previous iterative steps are used when parameters are estimated.

6.9.2 Example—First-order Model

The model is $\eta = L[1 - \exp(-kt)]$. There are $p = 2$ parameters and we will plan an experiment with $n = 2$ observations placed at locations that are optimal with respect to the best informed initial estimates of the parameter.

The derivatives are

$$X_{1u} = \partial[L(1 - e^{-kt}u)]/\partial L = 1 - e^{-kt}u, \qquad u = 1, 2$$

$$X_{2u} = \partial[L(1 - e^{-kt}u)]/\partial k = Lk\, e^{-kt}u, \qquad u = 1, 2$$

where t_1 and t_2 are the times at which observations will be made.

The objective now is to maximize the determinant of the $X'X$ matrix (max $\Delta = |X'X|$) constructed from these derivatives. In a complicated problem, the matrix multiplication and the minimization of the determinant of the matrix would be done using numerical methods. The analytical solution for this example (and quite a few other interesting models) can be derived rather easily. For $n = p = 2$, $\Delta = (X_{11}X_{22} - X_{12}X_{21})^2$. This is maximized when the absolute value of the quantity $(X_{11}X_{22} - X_{12}X_{21})$ is maximized. Thus, the design criterion is simply $\Delta^* = \max |X_{11}X_{22} - X_{21}X_{21}|$. Here the vertical bars designate the absolute value. The asterisk merely indicates the redefinition of the criterion Δ.

Substituting the appropriate derivative elements gives

$$\Delta = (1 - e^{-kt_1})(Lk\, e^{-kt_2}) - (1 - e^{-kt_2})(Lk\, e^{-kt_1})$$
$$= Lk[(1 - e^{-kt_1})(e^{-kt_2}) - (1 - e^{-kt_2})(e^{-kt_1})].$$

Lk is a constant and can be deleted. We only need to maximize the quantity in brackets.

The solution is $t_1 = 1/k$ and $t_2 = \infty$. The design criterion advises making measurements at t_2 as large as seems reasonable. This will provide a direct estimate of L, since as $t \to \infty$, $\eta \to L$, and the estimate will be essentially independent of k. The other observation at $t = 1/k$ is on the rising part of the curve. If we estimate $k = 0.23$ day^{-1}, $t_1 = 1/0.23$ days and observations should be placed at this location, say at 4, 4.5, or 5 days. Figure 6.8 shows how well this design worked on the BOD experiment; six well-

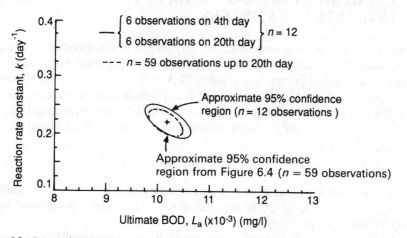

Figure 6.8. Approximate 95% joint confidence region for long-term BOD experiments using the $n = 59$ observations shown in Figure 6.4 and a subset of $n = 12$ observations (6 on day 4 and 6 on day 12, as suggested by the 'optimal' design)

placed observations gave virtually the same joint confidence region as 59 observations.

The efficiency of different designs can be compared using this criterion. For example, an experiment with five observations at days 3, 4, 5, 6, and 7 gives $\Delta^{1/2} = 353$; using five observations at days 2, 4, 6, 12, 14, 15 gives $\Delta^{1/2} = 871$. The efficiency of the latter experimental design, measured by the size of the joint confidence region, is approximately 250% greater than the first. These Δ values do not indicate the shape of the confidence region, but it happens that the best designs (largest $\Delta^{1/2}$) give well-conditioned (more ellipsoidal) regions because parameter correlation has been reduced.

6.9.3 Comments

It happens that the most efficient experimental strategy is to start with very simple designs, even as simple as $n = p$ observations, and then to work iteratively. Do a small experiment, estimate the parameters, use what has been learned about the shape of the curve to plan a few more experiments, and so on. This approach is illustrated after the basic principles have been outlined.

In many cases, the experimenters will not want to make measurements only at the p locations that appear optimal from maximizing Δ. They may prefer to use several observations in the region of the 'optimal' locations, for example, because setting up the experiment is costly but each measurement is inexpensive. They may want observations at other locations to check the adequacy of the model. This is sensible. The design criterion, after all, provides advice, not orders.

6.10 USING MULTI-RESPONSE DATA TO ESTIMATE PARAMETERS

Frequently data can be obtained simultaneously on two or more responses at a given experimental setting. Consider, for example, the reaction sequence where species A is converted to B which is in turn converted to C: $A \rightarrow B \rightarrow C$. The concentrations A, B and C of species A, B and C can be measured at any time of interest. Suppose that both steps of the reaction are first order and irreversible, with rate constraints k_1 and k_2. For $A = 1$ and $B = 0$ at time $t = 0$, the concentration of species B calculated at time t_1 is

$$B_{c,i} = [k_1/(k_2 - k_1)][e^{-k_1 t_i} - e^{-k_2 t_i}]$$

If only B is measured, k_1 and k_2 will be estimated by finding k_1 and k_2 to minimize $S = \sum(B_{o,i} - B_{c,i})^2$. The observed and calculated values are differentiated by the subscripts o and c, respectively.

If, in addition to B, both A and C are measured, better estimates of k_1 and k_2 can be obtained by using all the data and the appropriate models for A and C. A logical criterion for simultaneously fitting all three measured responses would be simply an extension of the non-linear least squares formulation, and this is to minimize the combined least squares for all three responses:

$$\min \sum (A_{c,i} - A_{o,i})^2 + \sum (B_{c,i} - B_{o,i})^2 + \sum (C_{c,i} - C_{o,i})^2$$

This criterion holds only if three fairly restrictive assumptions are satisfied, namely, (1) the errors of each response are normally distributed and all data points for a particular response are independent of one another, (2) the variances of all responses are equal, and (3) there is no correlation between data for each response for a particular experiment.

Assumption 2 is violated when certain responses are measured more precisely than others. This is probably more common than all responses being measured with equal precision. If assumption 2 alone is violated, it becomes appropriate to weight the sum of squares terms in inverse proportion to their variances. If both assumptions 2 and 3 are violated, the analysis must account for variances and covariances of the responses.

Box and Draper (1965) derived the criterion that is generally used. It is a combined sum of squares weighted by response. For the reaction $A \rightarrow B \rightarrow C$ it is

$$\min |V| = \begin{vmatrix} \sum (A_o - A_c)^2 & \sum (A_o - A_c)(B_o - B_c) & \sum (A_o - A_c)(C_o - C_c) \\ \sum (A_o - A_c)(B_o - B_c) & \sum (B_o - B_c)^2 & \sum (B_o - B_c)(C_o - C_c) \\ \sum (A_o - A_c)(C_o - C_c) & \sum (B_o - B_c)(C_o - C_c) & \sum (C_o - C_c)^2 \end{vmatrix}$$

Observed and calculated values are differentiated by subscripts o and c, respectively. The best parameter estimates are those that minimize the determinant of this matrix. Vertical lines indicate the determinant of the matrix.

The diagonal terms correspond to the sums of squares of each of the three responses. The off-diagonal terms account for measurements on the different responses being correlated. For the particular case of a single response this determinant criterion simplifies to the method of least squares: $\min \sum (A_o - A_c)^2$.

In general, the matrix is square with dimensions determined by the number of responses being fitted. Since the number of responses is usually small (say two to five), the above pattern for creating the matrix is easy to extend. Any non-linear optimization program should handle the minimization calculations.

The precision of the parameter estimates can be quantified by determining the approximate 95% Bayesian joint confidence region. The values of the determinant criterion $|V|$ take the place of the sum of squares function values in the usual single response estimation problem. The approximate 95% joint confidence region is bounded by

$$|V|_{1-\alpha} = |V|_{\min} \exp[\chi^2_{p(1-\alpha)}/n]$$

where $p =$ number of parameters estimated, $n =$ number of observations and $\chi^2_{p(1-\alpha)} =$ chi squared value for p degrees of freedom and the $(1 - \alpha)100\%$ probability level.

A comprehensive recent article of multi-response estimation, with several excellent discussions, is Bates and Watts (1985). Johnson and Berthouex (1975a, b) show an application to bacterial growth kinetics.

6.11 THE ITERATIVE APPROACH TO MODELLING

The dilemma of model building is that what needs to be known in order to design good experiments is what the experiments are supposed to discover. We could be easily frustrated by this if we imagined that we had to design one grand experiment that would lead directly to the desired results. Life, science, and statistics does not work this way. Knowledge is gained in small portions, each portion helping to guide the way to the next. Between each step there is need for reflection, study, and creative thinking. Experimental design, then, is a strategy as much as a technique.

An efficient experimental strategy is to start with very simple experimental designs, even as simple as $n = p$ observations. Do a small experiment, estimate the parameters, use what has been learned about the shape of the curve to plan a few more experiments, and so on.

6.11.1 Example—Bacterial Growth

Bacterial growth under steady state conditions in a completely mixed reactor is described by

$$X = Y(S_0 - S_i) \quad \text{and} \quad S = k/(FK/V - 1)$$

where F is the liquid flow rate, V is the reactor volume, S_0 is the influent substrate concentration, X and S are, respectively, the bacterial and substrate concentrations of the effluent.

A series of experiments will be done by changing the flow rate while keeping feed substrate concentration constant. A long period of operation is required for the chemostat to come to steady state after a new flow rate is set for the next experiment and, as a result, each observation is expensive. This situation suits the iterative approach ideally.

The three parameters will be estimated by fitting the two equations simultaneously using the multi-response criterion described above. Two experiments provide four data points (X_1 and S_1 at F_1; X_2 and S_2 at F_2) and this allows the three parameters to be estimated.

The optimal design advises doing an equation with $F = Vk/2$ and one at a flow rate as near the washout rate, $F_c = kS_0V/(K - S_0)$ as the experimenter dares. Measurements at $F > F_c$ are impossible because the bacteria are washed out of the reactor faster than they can grow. Also, conditions become unstable as F_c is approached. Working too close to F_c might result in a failed experiment, but staying too far on the safe side will not help much in estimating k.

The experimental design was to do only two experiments at each iteration. Figure 6.9 shows the settings of the independent variables, the data obtained, and the parameter estimates at each stage of the project. The 'data' were simulated using the true underlying values $Y = 0.6$, $k = 0.55$ and $K = 50$, and adding 'measurement error'.

Figure 6.10 shows the joint confidence regions obtained with this approach. Six experiments have given precise estimates of all three parameters, even though the

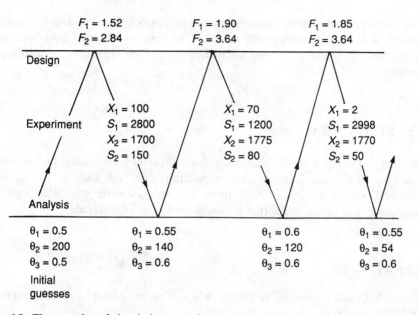

Figure 6.9. Three cycles of the design–experiment–analyse iterative approach to modelling. There were two experiments per cycle; the settings of the independent variable in each experiment was F and F. The observed values of the two dependent variables were (S_1, X_1) and (S_2, X_2). The resulting parameter estimates, computed using the multiresponse parameter estimation method, are shown

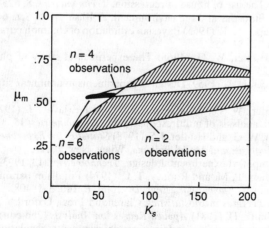

Figure 6.10. Approximate 95% joint confidence regions obtained at each of the iterative cycles of experimentation

initial guesses for parameter values were not particularly good. Y and k had already approached their true underlying values. The iterative experimental approach is very efficient. It is recommended particularly when measurements are difficult or expensive.

6.12 SUMMARY

Statistical methods for model building include parameter estimation and experimental design. Non-linear least squares is generally the best method for estimating parameters. Good designs provide more precise parameter estimates with fewer experiments. The iterative approach should always be considered.

REFERENCES

Atkinson, A. C. and Hunter, W. G. (1968). Design of experiments for parameter estimation, *Technometrics*, **10**, 271.

Bates, D. M. and Watts, D. G. (1985) Multiresponse estimation with special application to linear systems of differential equations, *Technometrics*, **27**, 329-39 (with discussion, 340-60).

Berthouex, P. M. and Hunter, W. G. (1971) Problems associated with planning BOD experiments, *J. San. Engr. Div., ASCE*, **97**, 333-444.

Berthouex, P. M. and Hunter, W. G. (1971) Statistical experimental design: BOD tests, *J. San. Engr. Div., ASCE*, **97**, 393-407.

Berthouex, P. M., Hunter, W. G., and Pallesen, L. (1981) Wastewater treatment: a review of statistical applications, in *ENVIRONMETRICS 81—Selected Papers*, pp 77-99, SIAM, Philadelphia.

Berthouex, P. M. and Scewczyk, J. E. (1984) Discussion of 'Influence of toxic metal on the repression of carbonaceous oxygen demand', *Water Res.*, **18**, 385-6.

Box, G. E. P. (1966) The use of abuse of regression, *Technometrics*, **8**, 625-9.

Box, G. E. P. (1974) Statistics and the environment, *J. Wash. Acad. Sci.*, **64**, 52-9.

Box, G. E. P. and Draper, N. R. (1965) Bayesian estimation of common parameters from several responses, *Biometrika*, **52**, 355.

Box, G. E. P. and Hunter, W. G. (1965) The experimental study of physical mechanisms, *Technometrics*, **7**, 23.

Box, G. E. P. and Lucas, H. L. (1959) Design of experiments in nonlinear situations, *Biometrika*, **45**, 77-90.

Box, G. E. P., Hunter, W. G., MacGregor, J. F., and Erjavac, J. (1973) Some problems associated with the analysis of multiresponse data, *Technometrics*, **15**, 33.

Box, G. E. P., Hunter, W. G. and Hunter, J. S. (1978) *Statistics for Experiments, An Introduction to Design, Data Analysis, and Model Building*, Wiley, New York.

Box, M. J. (1971) Simplified experimental design, *Technometrics*, **13**, 19-31.

Boyle, W. C., Berthouex, P. M. and Rooney, T. C. (1974) Pitfalls in parameters estimation for oxygen transfer data, *J. Envir. Engr., Engr. Div., ASCE*, **100**, 391-408.

Coloquhoun, (1971) *Lectures in Biostatistics*, Clarendon Press, Oxford.

Draper, N. R. and Smith, H. (1981) *Applied Regression Analysis* (2nd edn), Wiley, New York.

Hunter, J. S. (1977) Incorporating uncertainty into environmental regulations, in *Environmental Monitoring*, National Academy of Sciences, Washington, D.C.

Hunter, J. S. (1980) The national measurement system, *Science*, **210**, 869-74.

Hunter, J. S. (1981) Calibration and the straight line: current statistical practices, *J. Assoc. Anal. Chem.*, **64**, 574–83.

Hunter, W. G. (1982) Environmental statistics, in *Encyclopedia of Statistical Sciences*, Vol. 2, Eds. Klotz and Johnson, Wiley.

Johnson, D. B. and Berthouex, P. M. (1975a) Using multiresponse data to estimate biokinetic parameters, *Biotech. and Bioengr.*, **17**, 571–83.

Johnson, D. B. and Berthouex, P. M. (1975b) Efficient biokinetic design, *Biotech. and Bioengr.*, **18**, 557–70.

Joiner, B. L. (1981) Lurking variables: some examples, *Amer. Stat.*, **35**, 227–33.

Chapter 7

Models of Water Quality in Rivers

A. JAMES AND D. J. ELLIOTT

7.1 INTRODUCTION

Rivers have been used consistently as the principal pathway for disposing of wastewaters, industrial, domestic and agricultural. Clearly, rivers possess several attractive features as a means of wastewater disposal:

(a) conveying the wastewaters away to the sea;
(b) often rapid mixing which dilutes and disperses;
(c) often slow sedimentation and re-suspension which spreads the sediment over a large area;
(d) often turbulent conditions which causes rapid re-aeration.

However, despite these advantages there may be undesirable changes in flora and fauna. The majority of these changes are brought about by the discharge of organic matter (*BOD*) resulting in a decrease in the concentration of dissolved oxygen (*DO*).

Most river models are therefore concerned with the relationship between the rate of *BOD* discharge and the resulting concentration of *DO*. Various models have been used for this purpose, beginning with the classic work of Streeter and Phelps. Subsequent developments have produced a wide range of *DO* models, which are reviewed in the following pages.

From a systems analysis viewpoint the *DO–BOD* relationship is a particular example of a general situation which needs to be modelled.

Recent developments have included the formulation of generalized models which include a wide variety of water quality parameters and modular models which slot in the appropriate biochemical routines into a hydrodynamic structure.

An Introduction to Water Quality Modelling. 2nd Edition. Edited by A. James

In addition to extending the range of water quality parameters, recent developments in river models have also included two-dimensional models for mixing zones, stochastic models for compliance prediction and alternatives to the convective diffusion equation for representing mixing in streams.

The following notes review these developments after discussing the application of the convective diffusion equation.

7.2 CONVECTIVE DIFFUSION EQUATION

The convective diffusion equation is usually expressed in one-dimensional form as

$$\frac{\partial c}{\partial t} + U \frac{\partial c}{\partial x} = \frac{1}{A} \frac{\partial}{\partial x} \left(EA \frac{\partial c}{\partial x} \right) + La \tag{7.1}$$

where

La = source term

The derivation as discussed in Chapter 5 assumes an existing concentration of a conservative substance in the fluid. For most practical applications the equations must be extended to include source and sink terms for polluting discharges. If, for example, it is assumed that the material being discharged follows some predictable behaviour, e.g. it is conservative or undergoes a first-order decay, then equation (7.1) can also be extended to include both sources and sinks. But as the behaviour becomes more complex, it becomes more difficult to find an analytical solution and numerical methods are increasingly employed.

In the simple cases where the material is undergoing first-order decay equation (7.1) can be modified for steady-state situations to give

$$\frac{\partial c}{\partial t} + U \frac{\partial c}{\partial x} - \frac{1}{A} \frac{\partial}{\partial x} \left(EA \frac{\partial c}{\partial x} \right) = kC + La = 0$$

As shown in Figure 7.1 the source term may be approximated by a mass balance:

$$La = \frac{W}{Q + Qw}$$

where W = Mass flux discharged

Q and Qw are the freshwater flow and effluent flow respectively.

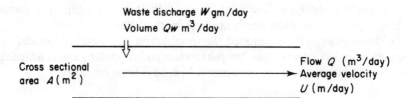

Figure 7.1. Conceptual diagram of a river model

7.2.1 Analytical Solutions

Particular solutions of equation (7.1) must satisfy the given differential equation and also comply with given initial and boundary conditions.

The initial conditions specify the values of La and C as functions of x along the stretch of interest at $t = 0$. The boundary conditions specify the values of L and C as functions of time at the beginning and end of the stretch. Boundaries must be chosen at sufficient distance from the stretch under consideration for the solution to closely approximate the specified conditions.

One form of analytical solution for equation (7.1) assumes that the discharge is continuous and the system has reached steady state, i.e., $\partial c/\partial t = 0$.

Also the stretch under consideration is assumed to have uniform flow and uniform cross-sectional area and the material being discharged does not significantly affect the flow in the river. Equation (7.1) now becomes

$$E \frac{\partial^2 c}{\partial x^2} - U \frac{\partial c}{\partial x} - Kc = 0 \tag{7.2}$$

assuming a first-order decay term.

The steady-state solution for this situation is given by

$$C = \frac{W}{AUm} \exp\left(\frac{U}{2E}[1 \pm m]x\right) \tag{7.3}$$

where

$$m = \sqrt{\left(1 + 4\frac{KE}{U^2}\right)}$$

In equation (7.3) the negative value of the exponent applies to the region downstream of the point of discharge. The positive exponent applies to upstream.

Changes in the relative magnitudes of E and K in equation (7.3) produce different answers as shown below.

In Figure 7.2, $K = 0$ (i.e. a conservative substance) the concentration at the discharge point is equal to W/AU, the rate of addition of the substance divided by the

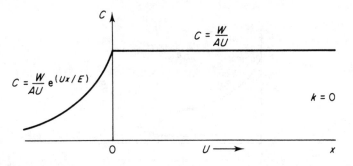

Figure 7.2. Concentration profile for $k = 0$

rate of flow in the river. The concentration is constant downstream of the outfall and decreases exponentially upstream.

In Figure 7.3, $E = 0$, no transport upstream occurs. Downstream the concentration decays from an initial value of W/AU at the point of discharge governed by

$$C = \frac{W}{AU} e - \left(\frac{Kx}{U}\right) \tag{7.4}$$

In Figure 7.4, both E and K are non-zero. The initial concentration is given by $C = W/AUm$ and decays both upstream and downstream. In practice because of the unidirectional nature of E there is no significant transport upstream of the point of discharge and the initial concentration is essentially equal to W/AU. This is because most dispersion is caused by velocity shear rather than turbulent eddies so that the upstream dispersion mechanism is much smaller than the downstream advection of bulk water flow.

Another form of particular solution to equation (7.1) is the situation where W is an instantaneous conservative discharge, again to a stream of uniform cross-sectional area.

The release at $t = 0$ and $x = 0$ produces a Gaussian concentration distribution with respect to x. The centre of the distribution moves downstream at velocity U as shown in Figure 7.4b.

The solution to this equation is

$$C = \frac{W}{A4\pi Et} \exp\left(\frac{x - Ut^2}{4Et}\right) \tag{7.5a}$$

where

W = weight of conservative substances
A = cross sectional area
t = time
x = distance downstream
U = mean velocity
E = dispersion coefficient.

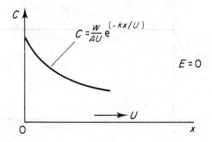

Figure 7.3. Concentration profile for $E = 0$

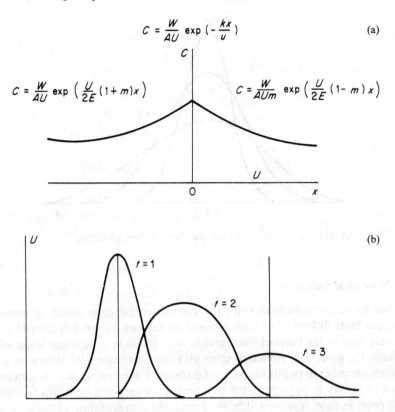

$$C = \frac{W}{AU} \exp\left(-\frac{kx}{u}\right) \qquad\qquad \text{(a)}$$

$$C = \frac{W}{AU} \exp\left(\frac{U}{2E}(1+m)x\right) \qquad\qquad C = \frac{W}{AUm} \exp\left(\frac{U}{2E}(1-m)x\right)$$

Figure 7.4. Concentration profiles for (a) non-zero values of E and k; (b) gulp injection

When there is no advection, e.g. if the discharge is into a canal, the solution is

$$C = \frac{W}{A4\pi Et} \exp\left(\frac{-x^2}{4Et}\right) \qquad\qquad (7.5b)$$

In this case the concentration profiles are those shown in Figure 7.5.

It is possible to estimate the value of E from tracer studies in which a slug of tracer is injected into the river. The time concentration curve of the tracer is measured at two stations downstream of the injection plot. E is obtained from

$$E = \frac{U^2}{2} \frac{\sigma_1^2 - \sigma_2^2}{t_2 - t_1} \qquad\qquad (7.6)$$

where

$t_1\ t_2$ are the mean times of the passage of the tracer past each station

σ_1^2 and σ_2^2 are the variances of the time concentration curves at stations 1 and 2

U is the mean velocity of flow between stations.

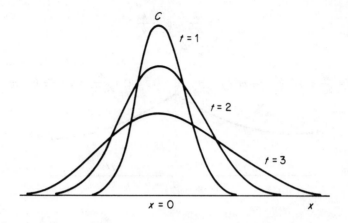

Figure 7.5. Concentration profile for a gulp injection in zero advection

7.2.2 Numerical Solutions

It is often more convenient to solve the convective diffusion equation numerically using either finite difference or finite element techniques. Finite differences have been extensively used as the basis of river models with a variety of schemes being adopted. Essentially, the numerical approach attempts to approximate the continuous solution at a discrete number of points in time and distance. For example, the one-dimensional equation (7.1) may be approximated on a time distance plane as shown in Figure 7.6. At each point in time, represented by the rows j, the concentration of the parameter of interest must be evaluated at each distance mesh point, represented by the columns i.

An example of this form of solution is demonstrated using an explicit finite difference scheme. The solution must satisfy certain initial and boundary conditions. At time $t = 0$ concentration levels must be specified at each distance mesh point. Also

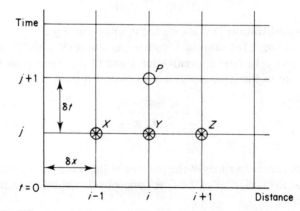

Figure 7.6. Finite difference scheme for solving the convective-diffusion equation

sufficient distance mesh points must be chosen so that the solution in the stretch of interest is not affected by the specified boundary conditions. Values at the boundaries are often chosen to represent natural river conditions.

In order to simplify the notation, equation (7.1) may be modified by assuming a constant cross-sectional area and a conservative pollutant to give

$$\frac{\partial c}{\partial t} = -U\frac{\partial c}{\partial u} + E\frac{\partial^2 c}{\partial x^2} + La \tag{7.7}$$

In explicit difference form equation (7.7) may be written as

$$\frac{C_{i,j+1} - C_{i,j}}{\delta t} = -\frac{-U}{\delta x}(C_{i-1,j} - C_{i,j}) + \frac{E}{\delta x^2}(C_{i-1,j} - 2C_{i,j} + C_{i+1,j}) + La \tag{7.8}$$

where La is the concentration which must be added to each mesh point at which a discharge takes place. The value of La is obtained from a mass balance of the effluent load and the river flow at the point of entry. It is convenient for the effluent discharge points and the mesh points to coincide. If not it is necessary to interpolate between adjacent mesh points.

Inspection of equation (7.8) shows some interesting features. Firstly let us assume dispersion is negligible in relation to advection. Thus if $E = 0$ equation (7.8) reduces to a purely advective equation. Considering Figure 7.7 it can be seen that for pure advection to be described correctly by the difference scheme the element of water at the point $i - 1$ at time j must move to i at time $j + 1$. Thus

$$C_{i,j+1} = C_{i-1,j} \tag{7.9}$$

For equation (7.8) to satisfy this requirement it is necessary that

$$U\,\delta t = \delta x \tag{7.10}$$

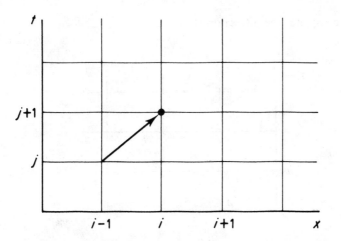

Figure 7.7. Finite difference scheme for pure advection

Secondly, if $U = 0$, then equation (7.8) is describing dispersion without convection. For this situation, a necessary condition for stability of the numerical solution is that

$$\frac{\delta t}{\delta x^2} E \leqslant \frac{1}{2} \qquad (7.11)$$

Assuming that the requirement of equation (7.10) is satisfied equation (7.8) may be simplified to

$$C_{i,j+1} = C_{i,j} + \frac{\delta t}{\delta x^2} E(C_{i-1,j} - 2C_{i,j} + C_{i+1,j}) \qquad (7.12)$$

It would now seem possible using equation (7.12) that a program could be written which calculates the unknown concentrations on the $j + 1$ row for all i columns using the known values of C on the j row. Inspection of equation (7.12) shows that if $E = 0$ then equation (7.9) is satisfied for pure convection. However, there is a problem with using equation (7.12) in its existing form.

Figure 7.8 shows a clean river into which effluent is being discharged. For a conservative substance under pure advection the concentration at B at time $j + 1$ must be the same as that at A at time j. Intuitively it can be seen that if dispersion processes are also taking place then the concentration at B must be less than at A. Equation (7.12) calculates a value for B ($C_{i,j+1}$) which is greater than at A ($C_{i-1,j}$). This results in a continuous build-up of concentration downstream of the discharge point which is artificially too high because of the numerical scheme used.

A two-step explicit method to overcome this problem is as follows:

(1)   advect the pollutant downstream for one time step.

$$C_{i,j} = C_{i-1,j} \quad \text{for all} \quad i$$

and

$$C_{0,n} = C_{0,j+1} \qquad (7.13)$$

This completes the advection step.

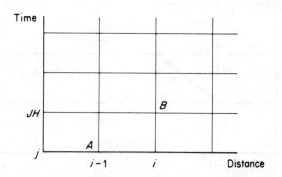

Figure 7.8. Finite difference scheme for pure advection of a conservative pollutant

(2) calculate new values on the $j + 1$ row using only the dispersion and loading terms. For example, the new concentration at P in Figure 7.9 is found using the values at X, Y and Z. The reduced equation for this step is

$$\frac{C_{i,j+1} - C_{i,j}}{\delta t} = \frac{E}{\delta x^2}(C_{i-1,j} - 2C_{i,j} + C_{i+1,j}) + La \qquad (7.14)$$

which may be arranged to give

$$C_{i,j+1} = C_{i,j} + E\frac{\delta t}{\delta x^2}(C_{i-1,j} - 2C_{i,j} + C_{i+1,j}) + \delta t\, La \qquad (7.15)$$

Starting with initial conditions specified on row j at each point i new concentrations may be found on the $j + 1$ row for all i. These now become the values used to calculate concentration at $j + 2$ etc. and the calculation can continue as long as necessary either to reach steady-state conditions or to display the river quality response characteristics to a time-varying discharge (see Figure 7.10).

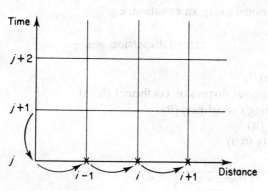

Figure 7.9. Two-step finite difference mesh

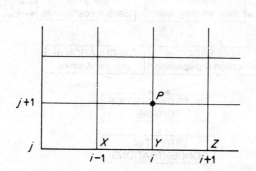

Figure 7.10. Calculation of pollutant concentration using two-step finite difference scheme

Recent developments in numerical methods, especially the QUICK routine, have considerably improved the stability and reduced the complexity of their computation (see Chapter 3 for details).

7.2.3 Applications of Finite Difference Models

Steady-state and dynamic models of water quality in rivers have been formulated using the finite difference schemes outlined above. The steady-state models have chiefly been used for setting consent conditions; the dynamic models have been used mainly for pollution incidents such as spills and stormwater discharges. The basic structure of this type of model, as illustrated in Figure 7.11, is based on the principle of the conservation of mass, both for the liquid and for any solutes. Basic hydraulic data in the form of:

(a) flow and velocity,
(b) flow and cross-sectional area or
(c) velocity and cross-sectional area

are generated from a hydrodynamic submodel using the principle of conservation of momentum or are obtained from field observations. Similarly dispersion may be calculated in the model using an equation, e.g.

$$\text{rate of dispersion} = E\,\frac{\partial^2 C}{\partial x^2}$$

where $E = 22.6\,n\,UD^{5/6}$
E = longitudinal dispersion coefficient (ft^2/s)
n = Manning's roughness (ft)
D = depth (ft)
U = velocity (ft/s)

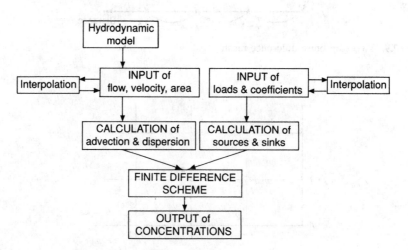

Figure 7.11. Flow diagram of a dynamic river model

or E may be calculated from other equations or may be specified in the Input data based on dye studies, or values published in the literature (see United States Environmental Protection Agency, 1984).

The interpolation subroutine indicated in Figure 7.11 serves to complete the matrices of values at all mesh points from the sparse data set that is usually available from field data. For example a model of a 30 km stretch of river may use a mesh size of 0.3 km and therefore have 101 mesh points. River surveys are unlikely to measure flow and velocities at more than 10 points, so the interpolation procedure is used to generate the remaining values, either in a linear form or in a stepwise manner, as shown in Figure 7.12.

Sources and sinks are best incorporated on an individual basis so that each module is self-contained and thus can be removed or amended without disturbing the other processes. Some typical ways of representing sources and sinks are indicated below, but see also United States Environmental Protection Agency (1985) for alternative representations.

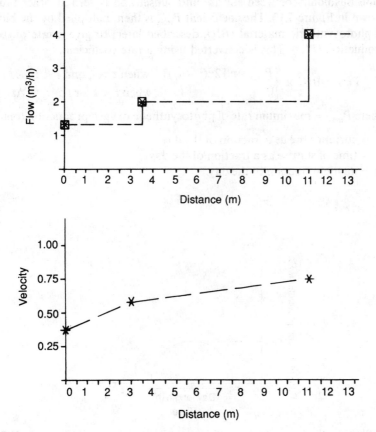

Figure 7.12. Interpolation routines

Each of the sources and sinks are described separately below.

(i) Surface reaeration (R) is considered, based on the empirical relationship for the coefficient AK_2

where

$$AK_2 = \frac{21.7\ U^{0.67}}{H^{1.85}}$$

with U = velocity (ft/s)

H = depth (ft)

and $R = AK_2\ (C_s - C)$

with R = rate of reaeration (mg/(l d))

C_s = saturation concentration of DO (mg/l)

C = current concentration of DO (mg/l)

All results are then transformed from imperial to metric units.

(ii) Photosynthesis is based upon a coefficient that varies with light intensity semisinusoidally between sunrise and sunset and is zero at other times as shown in Figure 7.13. The coefficient P_{max} is then multiplied by the biomass of photosynthetic material (BIO, described later) to give a rate of oxygen production ($P(t)$). This is converted using a rate coefficient.

$$P(t) = BIO \times \begin{cases} P_{max} \sin\left[2\pi(t - t_s)\right] & \text{when } t > t_s \text{ and } t < t_s + \Delta t \\ 0 & \text{when } t < t_s \text{ or } t > t_s + \Delta t \end{cases}$$

where P_{max} = maximum rate of photosynthetic oxygen production (mg/(l d))

t = current time as a fraction of the day

t_s = time of sunrise as a fraction of the day

Δt = daylight fraction of the day.

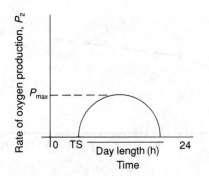

Figure 7.13. Photosynthetic oxygen production

(iii) The effect of reaeration by weirs (R) is incorporated using the relationship suggesting that R is proportional to weir height (H)

Hence $R = 1.0 \div \dfrac{(AW \times BW \times HW)}{2.0}$

where R = reaeration coefficient

$AW = \left. \begin{array}{l} 1.25 \text{ slightly polluted} \\ 1.00 \text{ moderately polluted} \\ 0.85 \text{ grossly polluted} \end{array} \right\}$ user selected

$BW = \left. \begin{array}{l} 1.3 \text{ cascade or stepped} \\ 1.0 \text{ free fall} \end{array} \right\}$ user selected

HW = height of weir (m)

The resulting coefficent (R) is then used in assessing the relationship between upstream and downstream deficit ratios.

(iv) Carbonaceous oxygen demand is calculated first by assessing the effect of settlement and resuspension, using the relationship

$$AKD = AK_1 + \frac{\cdot (AK_4 \times u)}{D}$$

with the rate of oxygen uptake (R_1) being calculated as

$$R_1 = BOD \times AKD \times A$$

where AK_4 = a resuspension coefficient, typically 0.2 (ranging from 0.1 to 0.6, and dependent on stream slope, as shown in Table 7.1)

AK_1 = a first-order BOD decay rate (h^{-1}) ranging from 0.006 to 0.015 but typically 0.01 (user provided)

U = velocity (= advection) (m^2)

A = cross-sectional area (m^2)

D = depth (m)

and AKD is a coefficient to be determined.

Table 7.1. The relationship between stream slope and the coefficient $AK4$

Stream slope %	AK_4 (dimensionless)
0.047	0.1
0.095	0.15
0.189	0.25
0.473	0.4
0.497	0.6

(After Bosko in United States Environmental Protection Agency (1985)).

(v) Benthal respiration (R_2) is calculated using a rate of respiration due to sediment oxygen demand which is dependent upon the coefficient *BRES*, entered by the user.

Hence, $R_2 = \dfrac{BRES \times \text{area}}{\text{depth}}$ (United States Environmental Protection Agency 1985)

and is expressed as g O m^2 h^{-1}

(vi) Plant respiration ($R3$) is calculated using the rate of respiration by plant and uses two inputted coefficients:

BIO = the biomass of plants per unit area (kg dry wt/m^2)
AK_5 = the rate of oxygen uptake per unit mass of plant biomass per hour (ranging from 0.5 to 2.0 but typically 1.5)

Hence $R_3 = \dfrac{BIO \times AK_5 \times \text{area}}{\text{depth}}$ (United States Environmental Protection Agency 1985)

(vii) Nitrification (i.e. the rate of oxidation of ammonia) is merely calculated as a first reaction dependent on a coefficient (AK_6) acting on the concentration of ammonia.

Hence $R_4 = AK_6 \times$ ammonia concentration $\times$ area (United States Environmental Protection Agency 1985)
where AK_6 = the ammonia decay rate (h), ranging from 0.005 to 0.05 and typically 0.02.

In the sink routines, R_4 is then multiplied by the stoichiometric coefficient relating mass of oxygen used, to the mass of ammonia oxidized (using the constant 4.33).

As mentioned previously, there are separate sink terms for each of the modelled parameters. In addition, temperature compensation and the calculation of oxygen deficits are corrected by inputting river water temperature (°C) and *DO* saturation concentration values (mg/l) for a range of temperatures, since the solubility of *DO* in water decreases with increasing temperature. Hence, all relevant coefficients (e.g. AK_1, AK_5, AK_6, P_{max}, *BRES*) are adjusted to a standard of 20°C in the forms listed below:

PW = temp $- 20.0$
AK_1 = $AK_1 \times 1.047\, PW$
AK_5 = $AK_5 \times 1.047\, PW$
AK_6 = $AK_6 \times 1.08\, PW$
P_{max} = $P_{max} \times 1.02\, PW$
$BRES$ = $BRES \times 1.047\, PW$

Where PW is the temperature correction to 20°C.

7.3 OXYGEN SAG MODELS

As mentioned in Section 7.1 the earliest attempts at a *DO/BOD* model involved the concept of oxygen sag. In this model the concentration of *DO* is represented as the resultant of two competing processes, namely:

(a) *Deoxygenation*—the rate of uptake of oxygen is regarded as being dependent on the remaining concentration of *BOD* and the rate coefficient K:

$$\frac{dDO}{dt} = -K_1 [BOD]$$

(b) *Reaeration*—the rate of replacement of oxygen is regarded as being dependent on the *DO* deficit and the rate coefficient K_2:

$$\frac{dDO}{dt} = K_2 [C_S - C]$$

where C_S = saturation *DO*
C = actual *DO*

Streeter and Phelps expressed the *DO* in terms of the deficit, D, giving the overall dynamics as

$$\frac{dD}{dt} = K_1 [BOD] - K_2 [D]$$

At the point of discharge the *BOD* will be high and the deficit will be low (assuming no upstream pollution), so the deficit will increase. Further downstream as the *BOD* is oxidized the rate of deoxygenation will decrease and the rate of reaeration will rise and then gradually fall as the deficit changes. These changes are illustrated graphically in Figure 7.14 and can be calculated using an integrated form of equation

$$Dt = \frac{K_1 [BOD]_0}{K_2 - K_1} (e^{-k_1 t} - e^{-k_2 t}) + [D]_0 e^{-k_2 t}$$

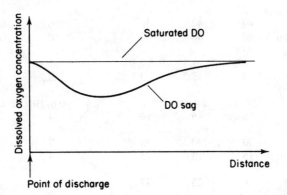

Figure 7.14. Oxygen sag curve

where t = time

$[BOD]_0$ and $[D]_0$ are the initial BOD and the DO deficit respectively.

This equation can be solved successively, using increasing values of t, to build up the deficit curve. Alternatively, if the only requirement is to calculate the minimum DO (critical point in Figure 7.14) then the following relationship can be used to calculate the value of t at the critical point:

$$t_c = \frac{1}{k_2 - k_1} \log_e \left[\frac{k_2}{k_1} \left(1 - \frac{[D]_0 (k_2 - k_1)}{k_1 [BOD]_0} \right) \right]$$

The greatest merit of the oxygen sag approach lies in the minimal requirement for field data, since the only requirements are as follows:

(i) *Initial BOD*—calculated from a mass balance on the discharge and the freshwater flow
(ii) *Initial deficit*—calculated from a similar mass balance
(iii) *Deoxygenation coefficient*—measured from successive oxygen consumptions in effluent-river water mixture
(iv) *Reaeration coefficient*—estimated by empirical equation or measured by gas tracer technique.

Clearly there are other processes in streams that may alter the DO and the BOD such as sedimentation, re-suspension, photosynthesis, etc., and in many streams these may be the principal sources and sinks as shown in Table 7.2. K_1 and K_2 are therefore

Table 7.2. Comparative rates of change of DO

| Process | Rate of change of DO (mg/l d) | | | | | |
| | Depth 0.5 m Velocity (m/s) | | | Depth 2.0 m Velocity (m/s) | | |
	0.03	0.15	0.3	0.03	0.3	0.3
Photosynthesis						
Macrophytes	40	40	40	10	10	10
Phytoplankton	40	40	40	10	10	10
BOD*						
20% effluent	1.3	1.3	1.3	1.3	1.3	1.3
50% effluent	3.2	3.2	3.2	3.2	3.2	3.2
Respiration						
Mud	2.4–43	2.4–20	0	0.6–10	0.4–7	0
Slimes	16	16	16	4	4	4
Respiration						
Macrophytes	30	30	30	7.5	7.5	7.5
Phytoplankton	15	15	15	3.7	3.7	3.7
Reaeration						
at 50% deficit	8	23	37	0.6	2	3

*This was calculated assuming that the effluent had a BOD of 20 mg/l.

lumped parameters which are more accurately calculated from stream *DO* and *BOD* data rather than from laboratory data or reaeration measurements. Various optimization techniques have been applied for this purpose which rely on a nonlinear, least squares technique. Usually the river is divided into a number of reaches, and within each reach K_1 and K_2 are held constant. Between reaches their values are not allowed to change by more than a maximum percentage, and the values of K_1 and K_2 are contained within acceptable ranges.

In its simplest form the oxygen sag model does not directly represent either advection or dispersion. Advection is used only in the sense of the relationship between time and distance travelled. Dispersion is included within K_1 and K_2.

However, more sophisticated versions have been formulated using the following equations:

$$\frac{\partial L}{\partial t} = -U\frac{\partial L}{\partial x} + E\frac{\partial^2 L}{\partial x^2} - (K_1 + K_3)L + La$$

$$\frac{\partial C}{\partial t} = -U\frac{\partial C}{\partial x} + E\frac{\partial^2 C}{\partial x^2} - K_1 L + K_2(C_s - C) - B$$

where L = first-stage *BOD*
La = rate of *BOD* discharged
K_3 = re-suspension coefficient
B = rate of oxygen uptake by benthal processes
E = coefficient of longitudinal dispersion
C = concentration of dissolved oxygen.

The above equations may be solved using a two-step explicit method, described in Chapter 3, first solving for *BOD* and then for *DO*. The dispersive and decay sub-equation may be written in finite difference form as

$$\frac{L_{i,j+1} - L_{i,j}}{\delta t} = \frac{E}{\delta x^2}(L_{i-1,j} - 2L_{i,j} + L_{i+1,j})$$

$$-\frac{K_1 + K_3}{2}(L_{i,j} + L_{i,j+1}) + La \tag{7.16}$$

and

$$\frac{C_{i,j+1} - C_{i,j}}{\delta t} = \frac{E}{\delta x^2}(C_{i-1,j} - 2C_{i,j} + C_{i+1,j})$$

$$-\frac{K_1}{2}(L_{i,j} + L_{i,j+1}) + \frac{K_2}{2}(C_s - C_{i,j}) - B \tag{7.17}$$

collecting terms in equation (7.16) and (7.17) gives

$$L_{i,j+1}\left(1 + \delta t\frac{(K_1 + K_3)}{2}\right) = L_{i-1,j}\frac{\delta t E}{\delta x^2} + L_{ij}\left(1 - \delta t\frac{(K_1 + K_3)}{2} - \frac{2\delta t E}{\delta x^2}\right)$$

$$+ L_{i+1,j}\frac{\delta t E}{\delta x^2} + \delta t La \tag{7.18}$$

and

$$C_{i,j+1}\left(1 + \frac{\delta t K_2}{2}\right) = C_{i-1,j}\left(\frac{\delta t E}{\delta x^2}\right) + C_{i,j}\left(1 - \frac{\delta t K_2}{2} - \frac{2\delta t E}{\delta x^2}\right)$$

$$+ C_{i+1,j}\left(\frac{\delta t E}{\delta x^2}\right) + \delta t K_2 C_s - \delta t B - (L_{i,j} + L_{i,j+1})\frac{\delta t K_1}{2} \quad (7.19)$$

A general program for the solution of equation (7.18) and (7.19) is given in the Appendix to this chapter.

The combination of the finite difference technique with the oxygen sag model makes a much more versatile formulation, which can be extended to include any desired range of process. The alternative approach that has been adopted is to modify the analytical solution to include additional processes. For example, the following formulation includes re-suspension, nitrification and photosynthesis as well as various forms of respiration.

$$D_{x,t} = D_0 e^{-K_a X/U} \qquad \text{initial deficit}$$

$$+ \frac{K_d L_0}{K_a - K_r}(e^{-K_r X/U} - e^{-K_a X/U}) \qquad \text{BOD including re-suspension}$$

$$+ \frac{K_n N_0}{K_a - K_r}(e^{-K_n X/U} - e^{-K_a X/U}) \qquad \text{nitrification}$$

$$+ \frac{R}{K_a}[1 - e^{-K_a X/U}] \qquad \text{mud respiration}$$

$$+ \frac{R}{R_a}\left[\frac{2p}{K_a}(1 - e^{-K_a X/U})\right] \qquad \text{plant respiration}$$

$$+ 2\sum \frac{a_n}{\sqrt{[(K_a)^2 + (2\pi_a)^2]}} \cos\left[2\pi_n\left(t - \frac{p}{2}\right) - tn^{-1}\left(\frac{2\pi_n}{K_a}\right)\right] \qquad \text{algal respiration}$$

$$P_m\left\{-2e^{-K_a X/U}\sum \frac{a_n}{\sqrt{[(K_a)^2 + (2\pi_a)^2]}} \cos\left[2\pi_n\left(t - \frac{p}{2}\right) - tn^{-1}\left(\frac{2\pi_a}{K_a}\right)\right]\right\}$$

$$\text{algal photosynthesis}$$

7.4 LAGRANGIAN MODELS

The finite difference approach uses a fixed Eulerian grid for the solution of the convective diffusion equation. An alternative approach is to use a Lagrangian formulation for the model. This approach assumes the observer is travelling at the same speed as the parcel of water under observation. If it is also assumed that the dispersive mechanism is negligible then biochemical decay is the only source of change within the box. This type of model tracks given parcels of water through time as they move downstream and calculates the rate of biochemical reaction at each increment. If the position of all the parcels is known at any one time and their concentrations have been calculated, then a one-dimensional plot of concentration against distance can be made for the parameter of interest.

7.4.1 Moving Segment Model

The basic idea is to simulate the flow in a stream as a series of blocks moving consecutively downstream. Within each block the variations in DO are calculated by summing hourly the changes due to all the processes involved as shown in Figure 7.15.

The rate of movement of the blocks is related to the flow rate and the cross-sectional area. As the flow varies the cross-sectional area is adjusted and a new time of travel is calculated.

The blocks are considered to be discrete and no change of DO or other materials is allowed between the blocks. This is not a serious disadvantage because the time step for segmentation is so short that adjacent blocks are unlikely to contain widely differing concentration of oxygen or organic matter.

As the blocks move downstream the rate of oxygen consumption and production varies due to differences in velocity, nature of the bed, organic concentration, daylight, etc. Some of these differences are associated with the changing physical characteristics of the river and for this reason the model river is divided into reaches. Within each reach the rates of reaeration and benthal respiration are regarded as uniform and the biomass of the rooted flora is considered to be evenly distributed.

The end of a reach marks a point of discontinuity in the physical conditions or may mark a discontinuity in the oxygen profile due to a weir. Subdivision of the river into reaches is also made at places where any major tributary or effluent water enter.

The mathematical representation of the oxygen concentration in a block may be summarized as follows:

$$\frac{dC}{dt} = P \pm RA - RE \tag{7.20}$$

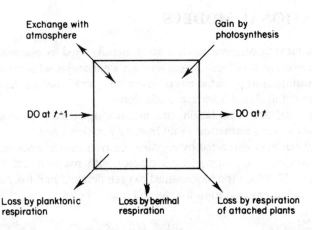

Figure 7.15. Mass balance on DO in a moving segment

where

$$P = \frac{B_1 \times Pmx}{K} \times \frac{I \times SA}{IK} + B_2 \times Pmx \frac{I}{Ik} \qquad (7.21)$$

$$RA = K_2 (C_S - C) \qquad (7.22)$$

$$RE = K_1(BOD) \times SA \qquad (7.23)$$

and

$$C_t = C_0 \int_0^t \frac{dC}{dt}$$

P = photosynthesis
RA = reaeration
RE = respiration
I = light intensity
IK = light intensity corresponding to Pmx
SA = surface area
B_1 = biomass of attached plants and algae
B_2 = biomass of planktonic algae
K = extinction coefficient for light
K_3 = respiration rate of bottom deposits
Pmx = maximum rate of photosynthesis
C = concentration of dissolved oxygen.

The inputs to the model define the upstream boundary conditions, the values of all the rate coefficients for each reach and the climatic conditions in terms of light intensity and temperature. Once one block has been followed through all reaches, then the program loops back to follow the oxygen concentration in the next block (not shown in Figure 7.16).

7.5 OPERATIONAL MODELS

Dispersion and moving segment models are generally used as planning tools, since they predict the average dissolved oxygen level for a particular set of conditions. There are some circumstances in polluted rivers where an operational model is required to give much more detailed and accurate predictions.

This involves real-time simulation and monitoring usually in connection with control of a river water abstraction or an in-river aeration system.

Such a model can be constructed by dividing the river into a series of reaches, each of which is represented as a stirred tank reactor with reaction and time delay, as shown in Figure 7.17. When the biochemical oxygen demand and the reaction are the main sink and source, the equation for the DO is as follows:

$$V \frac{dC_2}{dt} = Q(C_{1,t-d} - C_{2,t}) + VK_2(C_S - C_S - C_{2,t}) - VK_1C_{2,t} \qquad (7.24)$$

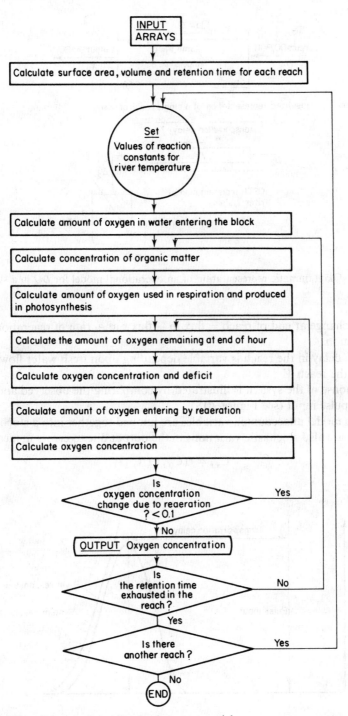

Figure 7.16. Flow diagram of the dissolved oxygen model

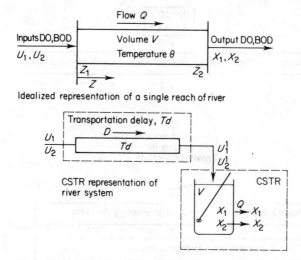

Figure 7.17. Diagrammatic representation of an operational model for *DO* in a stream

Rate of change at end of reach = flux in − flux out + rate of reaeration − rate of deoxygenation.

The time delay in the reach is variable depending upon fresh water flow Q and the volume of the reach V.

The response of the system is illustrated by comparing the observed and predicted data for a pulse input (see Figure 7.18).

In such a model unaccounted variables are lumped together into a stochastic input $R_{t,2}$. There are also stochastic variations in the output due to errors in measurement.

$$C_{n,t} = f(C_{0,t}), (N_{t,2}) \qquad (7.25)$$

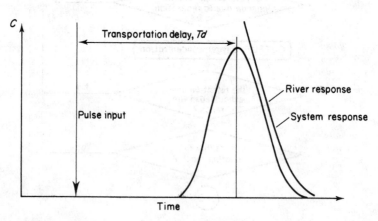

Figure 7.18. Comparison of river and assumed system response

Hence the full model for the reach can be written as

$$V \frac{dC_2}{dt} = Q(C_{1,t-d} - C_{2,t}) + K_2(C_s - C_{2,t}) - K_1 C_{2,t} + R_{t,2} \qquad (7.26)$$

where

$$C_{2,t} = C_{1,t} \pm N_{t,2}$$

As $(t, 2)$ and (t) have different frequencies they can be separated by time series analysis (see Chapter 4 for a description of techniques). For example, if $R(t, z)$ is a low-frequency stochastic term due to the omission of some slowly varying sink and $N(t, 2)$ is a random variable applied to the DO measurements, which are made at frequent intervals, then separation could be achieved by time averaging over a number of DO observations, since $R(t, 2) = 0$. In practice much more complex statistical techniques are employed, notably the Kalman filter.

Since $N(t, z)$ and $R(t, z)$ can be separated, the state parameters K_1 and K_2 can be evaluated and constantly updated by recursive estimates on observed data.

7.6 OPTIMIZATION MODEL

River models based on dispersion or moving segments can be used to generate a matrix of transfer coefficients by running the model several times with unit changes in BOD loadings. Thus, for a simple situation with two reaches and two waste treatment plants as shown in Figure 7.19, the DO/BOD relationship may be represented by a 2×2 matrix of coefficients. Each transfer coefficient ϕ represents the change in BOD loading from treatment in plant j.

$$\begin{bmatrix} \phi_{11} & \phi_{12} \\ \phi_{21} & \phi_{22} \end{bmatrix}$$

Given the data of the cost of treatment at two plants C_1 and C_2 (unit costs per kg BOD removed per day) and the improvement in DO required D_1 and D_2 (g O_2 per m^3) then the problem can be solved by a suitable optimization routine.

The objective function can be written as

minimize $$C = \sum_{j=1}^{n} C_j W_j$$

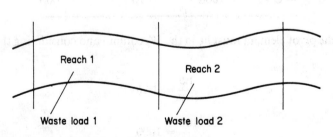

Figure 7.19. River system for optimization example

where

C = total cost

W_j = number of kg of *BOD* removal per day at treatment plant j

C_j = cost per kg of *BOD* removed per day at treatment plant j

and the constraint equation becomes

$$W_1 \times \phi_{11} + W_2 \times \phi_{21} \geq D_1$$
$$W_1 \times \phi_{12} + W_2 \times \phi_{22} \geq D_2$$

giving the following data

$D_1 = 1$
$D_2 = 2$
$W_1 = 10$
$W_2 = 20$

$$\begin{bmatrix} \phi_{11} & \phi_{12} \\ \phi_{21} & \phi_{22} \end{bmatrix} = \begin{bmatrix} 0.01 & 0.001 \\ 0.005 & 0.01 \end{bmatrix}$$

the problem may be solved by the simplex method.

	R_1	W_2	
R_1	0.01	0.001	1
R_2	0.005	0.01	2
$-C$	1000	10	-1000

The pivot element must lie in the W_2 column and the ratios 2/0.005 and 1/0.01 show that it must be the element in the top row. The tableau can therefore be transformed using the simplex method and the new tableau therefore becomes

	R_1	W_2	
W_1	100	0.1	100
R_2	-0.5	0.09	1.5
$-C$	1000	10	-1000

In this case the pivot element must lie in the W_2 column and considering the constraint ratios:

$$\frac{1.5}{0.09} = 16.6$$

$$\frac{100}{0.1} = 1000$$

it must be element R_2/W_2. Transforming the tableau about the element gives

	R_1		
		R_2	
W_1	105.3	10.5	85
W_2	− 52.5	105	157.5
− C	0	− 11445	− 4000

which is the final solution, indicating that the improvement in *DO* can be obtained by an additional 85 kg per day removal at treatment plant 1 and an additional 157 kg per day removal at treatment plant 2. The cost will be 4000 and the improvement in *DO* will be:

$$\text{Reach 1} \quad 85 \times 0.01 + 157.5 \times 0.001 = 1 \text{ mg/l}$$

$$\text{Reach 2} \quad 85 \times 0.01 + 157.5 \times 0.005 = 2 \text{ mg/l}$$

7.7 MODELS OF DISCHARGE

A model for a point of discharge has already been described in Chapter 1. As shown in Figure 1.1, the concentration of a conservative pollutant at the end of the mixing zone can be calculated from a simple mass balance:

$$C_m = \frac{Q_1 \times C_1 + Q_2 \times C_2}{Q_1 + Q_2}$$

where

C_m = concentration at end of mixing zone
C_1 and Q_1 = concentration and flow in fresh water
C_2 and Q_2 = concentration and flow in effluent

When the discharge is from a diffuse source the representation can be made by dividing the river into reaches so that within each reach the discharge is uniform over its length (see Figure 7.20)

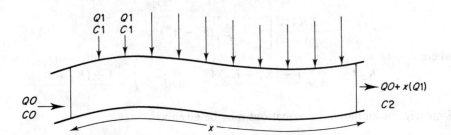

Figure 7.20. Conceptual model of a diffuse source

The mass balance on the conservative pollutant is given by

$$C_2 = \frac{Q_0 C_0 + \displaystyle\int_0^x Q_1 C_1 \, dx}{Q_0 + \displaystyle\int_0^x Q_1 \, dx}$$

Q_0 and C_0 are flow concentration upstream

   x is the length of the reach
   Q_1 is additional run-off per unit length
   C_1 is the concentration in the run-off
   C_2 is the concentration at the end of the reach

this integrates to give

$$C_2 = C_0 F + C_1(1 - F)$$

where

$$F = \text{dilution factor}$$

$$= \frac{Q_0}{Q_0 + Q_1 x}$$

As with point discharges, it is assumed that the pollutant mixes completely with the river water before the end of the reach.

For non-conservative substances, the same approach may be used suitably modified to take account of the chemical and/or biological processes that alter the concentration of the substance concerned.

For DO the appropriate equation is as follows:

$$\frac{d}{dx}(Q_0 + Q_1 x)d = Q_1 r_D + K_2 A(d_s - d) - K_1 Az - K_3 Aa$$

which may be integrated to give

$$d\,(L) = d_0 F^\delta + \frac{1}{q\delta}\left[r_D Q_1 + K_2 A d_s - \frac{K_1 A r_2}{\alpha} - \frac{EK_3 A r_A}{B} \right]$$

$$\times (1 - F^\beta) - \frac{K_1 A}{q(\delta - \alpha)}\left[z_0 - \frac{r_2}{\alpha} \right](F^\alpha - F^\delta)$$

$$- \frac{EK_3 A}{q(\delta - \beta)}\left[\alpha_0 - \frac{r_A}{\beta} \right](F^\beta - F^\delta)$$

where

$$\delta = 1 + \frac{K_2 A}{q} \qquad \alpha = \frac{1 + (K_1 + K_5)A}{q} \qquad \text{and} \qquad \beta = \frac{1 + (K_3 + K_4)A}{q}$$

Similarly, for ammonia the equations may be given as

$$\frac{d}{dx}(Q + qx)a = qr_A - (K_3 + K_4)A_a$$

and

$$a(L) = a_0 \, F^\beta + \frac{r_A}{\beta}(1 - F^\beta)$$

and other symbols are defined as follows:

A = average cross-sectional area
a = concentration of ammoniacal nitrogen
d = concentration of dissolved oxygen
E = relative oxygen usage in nitrification
r_D = DO in run off
r_A = ammoniacal nitrogen concentration in run off
r_z = BOD in run off
d_s = DO saturation
K_1 = BOD decay coefficient
K_2 = reaeration coefficient
K_3 = nitrification coefficient
K_4 = volatilization coefficient for ammonia
K_5 = BOD sedimentation coefficient

7.8 STREAM TUBE MODEL

The vast majority of discharges to rivers occur through pipes or channels which end at the bankside, so that in all but the smallest of streams, there is a distinct mixing zone in which water quality is not laterally uniform (see Figure 7.21). In some pollution situations it is important to be able to calculate the concentration distribution across the stream, especially in large rivers, since the length of the mixing zone (usually 50–100 × the width) can be very significant. For this reason mathematical techniques based on the stream tube concept have been developed.

The stream tube concept considers the flow in a river to be made up of a number of equal subflows. Longitudinal dispersion and near field effects are assumed to be negligible. It is assumed that there are no vertical differences in concentration and no density differences between the effluent and the river water. The steady-state distribution for a conservative substance is given by

$$\frac{\partial C}{\partial x} = \frac{\partial^2 C}{\partial q^2} - \frac{M_x C}{U}$$

where

C = concentration
x = distance longitudinally
q = cumulative discharge measured from a reference bank
M_x = scaling factor to adjust for the length of the tubes
U = depth averaged velocity

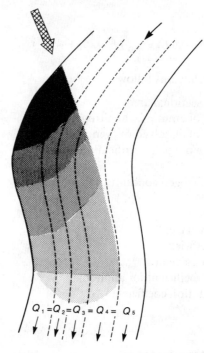

Figure 7.21. Stream tube model

The cumulative discharge, q, is defined by

$$q = \int_0^y (Muh)\,\mathrm{d}y$$

where

$\quad h\;\;$ = local depth at lateral distance y

$\quad M_y$ = scaling factor to adjust for the width of the tubes

The lateral diffusion coefficient D_y is assumed to be constant at a cross-section and is given by the average product of M_x, E_y, U and h^2

$$D_y = \overline{M_x E_y U h^2} = \frac{1}{Q}\int_0^Q (M_x E_y U h^2)\,\mathrm{d}q$$

where Q = discharge

$\quad\;\; E$ = lateral dispersion coefficient

Alternatively D_y can be expressed using a shape factor

$$F = \int_0^1 \left(\frac{h}{\bar{h}}\right)^3 \left(\frac{u}{\bar{u}}\right)^2 \mathrm{d}\left(\frac{y}{b}\right)$$

where y is the top width of the channel.

The shape factor F is to take account of the local deviations in depth and velocity, and has values between 1.0 and 3.2.

The lateral dispersion coefficient E_y can be expressed by

$$E_y = \beta_e b \bar{u}$$

where β is a dimensionless factor, so

$$D_y = M_x F \beta_e b \bar{u} h^{-2} = \frac{\beta Q^2}{b}$$

where β is a non-dimensional diffusion factor in the range 10^{-4} to 10^{-3}. The diffusion factor may be estimated from the variance

$$\delta q^2 = 2 D_y x$$

$$\frac{\delta q^2}{Q^2} = 2\beta \left(\frac{x}{6}\right)$$

in which δq^2 is the variance of C against q relationship.

7.9   STOCHASTIC MODELS

Compliance with water quality standards is increasingly being expressed in terms of probability. It is therefore important in simulating the quality of river water, to be able to include random variations in quality brought about by fluctuations in load, flow and other environmental variables. Two fundamentally different approaches have been adopted, which are:

(i)   steady-state Monte Carlo simulation;
(ii) dynamic time series simulation.

These are outlined briefly below.

7.9.1   Monte Carlo Simulation

In its simplest and most popular form this involves using a mass balance model repeatedly with varying input conditions to generate an output in the form of a frequency distribution of pollutant concentration. The methodology, as illustrated in Figure 7.22, consists of the following steps:

(i)   sample from a frequency distribution of flows;
(ii)   sample from a frequency distribution of loads;
(iii) calculate concentration after initial dilution;
(iv)   route the pollution downstream and calculate the resulting concentration allowing for decay, increased dilution, etc.;
(v)   construct the frequency distribution of concentrations resulting from repeating steps (i)–(iv).

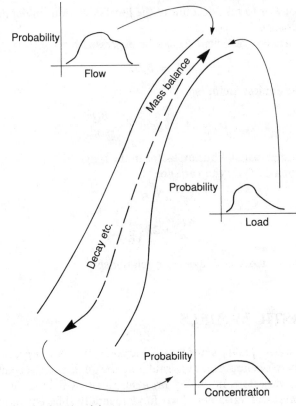

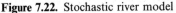

Figure 7.22. Stochastic river model

The important advantages of this approach are the minimal requirements for field data and computation. The latter allows the simulation to be run several hundred times, giving a statistically robust picture of the resulting distribution.

There are, however, two significant disadvantages. The first is due to the simplistic routing of pollution downstream, which is usually done using time-of-travel data. Processes like photosynthesis may cause marked diurnal variations in *DO* patterns.

These are considered as noise in the model, which makes it more difficult to distinguish the effect of different pollution strategies.

Secondly the model can be used to calculate percentage compliance but gives no indication of the frequency or duration of any lapses. This makes it difficult to interpret their ecological impact. Also the statistical basis of those models often assume that loads, flows, etc. are independent, random variables, whereas in reality they are strongly correlated and contain a high degree of autocorrelation.

7.9.2 Synthetic Time Series Model

This approach has been implemented using a dynamic finite difference model as described in Section 7.2. River data are analysed in two ways (see Figure 7.23):

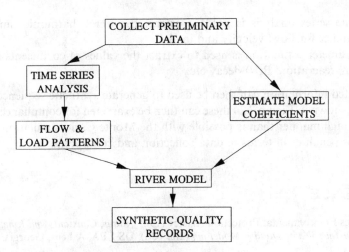

Figure 7.23. Flow diagram of data processing

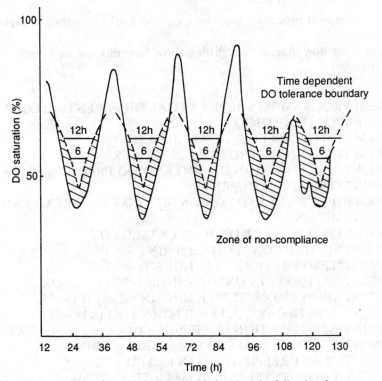

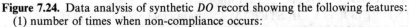

Figure 7.24. Data analysis of synthetic *DO* record showing the following features:
(1) number of times when non-compliance occurs:
(2) antecedent conditions prior to non-compliance;
(3) length of time for each non-compliance episode

(a) Time series analysis is used to extract the trend, harmonics and residual variance for flow, velocity and loads.

(b) Parameter estimation is used to extract the values of coefficients for dispersion, reaeration, *BOD* decay etc.

The extracted information can then be used to generate synthetic sequences of water quality. As shown in Figure 7.24 these can then be examined for compliance in a much more thorough manner than is possible with the Monte Carlo simulation. But there are obvious penalties in terms of data collection and computing.

REFERENCE

United States Environmental Protection Agency (1985) *Rates, Constants and Kinetic Formulations in Surface Water Quality Modelling* (2nd edn), US EPA, Athens, Georgia.

APPENDIX A

The two-step explicit procedure for computing *DO/BOD* in a stream is explained in Section 7.3.

The following flow diagram and listing show how this can be transformed into computer code.

```
10   REM PROGRAM TO SOLVE PARTIAL DIFFERENTIAL EQUATIONS
     BY EXPLICIT METHOD
20   REM FINITE DIFFERENCE APPROACH
30   REM BOD AND DO PROFILES IN A RIVER
40   PRINT "CALCULATION OF BOD AND DO PROFILES IN A RIVER"
50   PRINTSTRING$(75,"*"):PRINT
60   REM INPUT YOUR INITIAL CONDITIONS USING READ AND DATA
     STATEMENTS
70   READ T,SOX,E,K1,K3,B,D,R,P,V,D,LR,TM,DX,DT
80   PRINT"INITIAL CONDITIONS:":PRINT
90   PRINT"TEMPERATURE IN CELSIUS(T) = ";T
100  PRINT"SATURATION OXYGEN IN mg/l (SOX) = "; SOX
110  PRINT"DISPERSION COEFFICIENT IN m2/day (E) = ";E
120  PRINT"BOD DECAY COEFFICIENT IN day-1 (K1) = "; K1
130  PRINT"NITRIFICATION COEFFICIENT IN day-1 (K3) = ";K3
140  PRINT"BENTHAL DEMAND IN mg/l (BD) = ";BD
150  PRINT"PLANT RESPIRATION IN mg/l (R) = ";R
160  PRINT"PHOTOSYNTHESIS IN mg/l = ";P
170  PRINT"VELOCITY OF STREAM IN m/day (V) = "; V
180  PRINT"STREAM DEPTH IN m(D) = ";D
190  PRINT"STREAM WIDTH IN m (W) = ";W
```

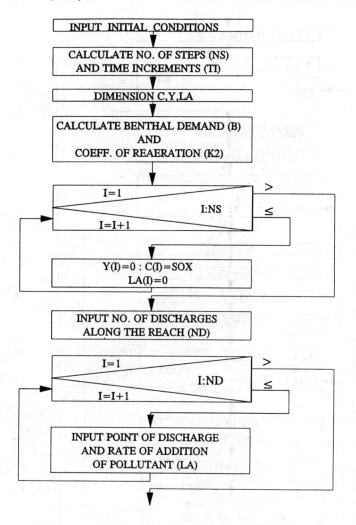

200 PRINT"LENGTH OF REACH IN m (LR) = ";LR
210 PRINT"TIME OF DISPERSION OF POLLUTANT IN days (TM) = ";TM
220 PRINT"DELTA Z IN m (DX) = ";DX
230 PRINT"DELTA T IN days (DT) = ";DT
240 PRINT:PRINT:PRINT
245 REM =
250 REM DEFINE NO. OF STEPS(NS) AND NO. OF TIME
     INCREMENTS(TI)
260 @% = &20209
270 NS = LR/DX
280 TI = TM/DT
285 REM =

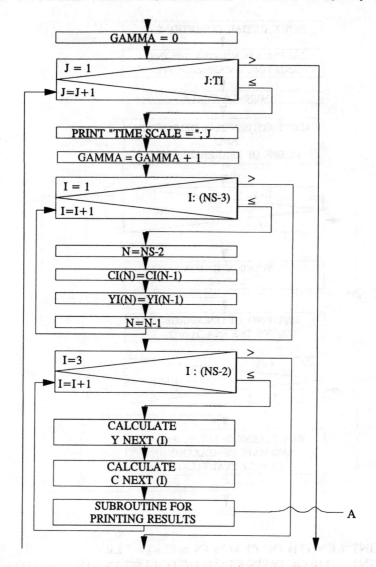

290 REM DIMENSION VARIABLES
300 DIM YI(100),YNEXT(100)
310 DIM CI(100),CNEXT(100)
320 DIM LA(100)
325 REM =
330 REM CALCULATE B AND K2 VALUES
340 B = BD + R − P
350 K2 = 9; 4*(V^0.67)*D^(− 1.85)
355 REM =

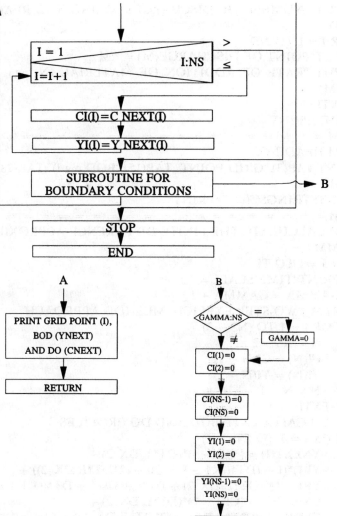

360 REM LOOP TO INITIALIZE
370 FOR I = 1 TO NS
380   YI(I) = 0
390   CI(I) = SOX
400   LA(I) = 0
410 NEXTI
420 REM =
430 REM CALCULATE THE RATE OF ADDITION OF MATERIAL AT THE POINT OF DISCHARGE

440   INPUT "NUMBER OF DISCHARGES ALONG THE REACH (ND) = ";ND

450   FOR I = 1 TO ND

460   INPUT"POINT OF DISCHARGE(M) = ";M

470   INPUT "RATE OF ADDITION OF MATERIAL IN mg/(l d)(LA) = "; LA(M)

480   NEXTI

490   PRINT:PRINT

500   REM =

510   REM HEADINGS

520   PRINT TAB(3);"GRID POINT";TAB(15);"BOD(mg/l)";TAB(28);"DO (mg/l)"

530   PRINTSTRING$(50,"*"):PRINT

540   REM =

550   REM CALCULATE THE FINITE DIFFERENCE APPROXIMATION

560   GAMMA = 0

570   FOR J = 1 TO TI

580   PRINT"TIME SCALE = ";J

590   GAMMA = GAMMA + 1

600   REM TWO-STEP EXPLICIT METHOD APPROACH

610   FOR I = 1 TO (NS − 3)

620   N = NS − 2

630   CI(N) = CI(N − 1)

640   YI(N) = YI(N − 1)

650   N = N − 1

660   NEXTI

670   REM CALCULATE BOD AND DO PROFILES

680   FOR I = 1 TO (NS − 2)

690   YNEXT(I) = (YI(I − 1)*(DT*E/DX^2) +
             YI(I)*(I − (DT*((K1 + K3)/2)) − (2*DT*E/DX^2)) +
             YI(I + 1)*(DT*E/DX^2) + DT*LA(I))/(1 + DT*((K1 + K3)/2))

700   CNEXT(I) = (CI(I − 1)*(DT*E/DX^2) +
             CI(I)*(1 − ((DT*K2)/2) − (2*DT*E/DX^2)) + CI(I + 1)*(DT*E/
             DX^2) + DT*K2*SOX − DT*B((YI(I) + YNEXT(I))*(DT*K1/2)))/
             (1 + (DT*K2/2))

800   GOSUB 1500

810   NEXTI

820   REM ITERATION PROCESS

830   FOR I = 1 TO NS

840   CI(I) = CNEXT(I)

850   YI(I) = YNEXT(I)

860   NEXTI

870   GOSUB 1600

880   NEXTJ

890   END

```
1500   REM = = = = = = = = = = = = = = = = = = = = = = = = = =
1510   REM PRINTING RESULTS
1520   PRINT TAB(3);I;TAB(15);YNEXT(I);TAB(28);CNEXT(I)
1530   RETURN
1600   REM = = = = = = = = = = = = = = = = = = = = = = = = = =
1610   REM BOUNDARY CONDITIONS
1620   CI(1) = 0:CI(2) = 0
1630   CI(NS − 1) = 0:CI(NS) = 0
1640   YI(I) = 0:YI(2) = 0
1650   YI(NS − 1) = 0:YI(NS) = 0
1660   RETURN
1700   REM = = = = = = = = = = = = = = = = = = = = = = = = = =
2000   DATA20,8.5,342700,0.25,0.15,0,0,0,27000,0.6,10,15000,0.6,500,0.05
```

APPENDIX B

The fundamental structure of finite difference models, of water quality in streams, requires relatively minor modification to enable a wide range of parameters to be computed. Some examples of subroutines are given in the appendix to Chapter 5, but further examples may help to clarify this approach. The following notes and flow diagram illustrate the formulation of a model to simulate the temperature distribution in a stream.

SORCT calculates the heat sources which include the following:

(a) heat loads from SORCT;
(b) solar radiation from SOLRAD;
(c) atmospheric radiation from ATRAD.

SINKT calculates the heat sinks which consist of the following:

(a) heat radiating back into the atmosphere BACRAD;
(b) heat losses due to evaporation EVAP;
(c) heat losses/gains due to temperature difference between stream water and atmosphere CONVEC.

C

```
       SUBROUTINE SORCT(T,I,NTPL,ATEMP1)
       COMMON /VEL/U(101),Q(101),A(101),D(101)
       COMMON /MESH/DX,DT,UMAX
       COMMON /TMM/DISTT(20),TEMP(20),TEMPT(101,60)
       COMMON /TML/DISTTL(20),TEMPL(101,50),TTTEMP(20,50),
       TL(20,101)
       COMMON /TPL/STEMP(101),TAL(101),TPAL(101)
       COMMON /MAX/CA,ALMAX,CB,QS
       COMMON /TEM/NTEMP,NTEMPL
       COMMON /TIM/SUNR,DAYLEN
```

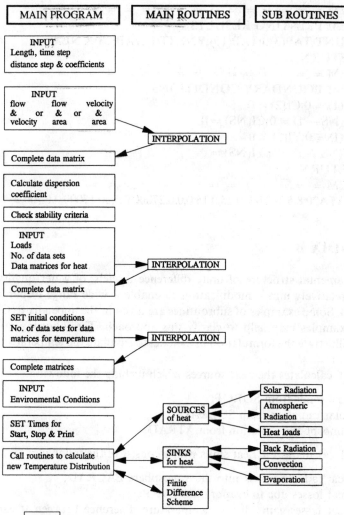

```
4           COMMON /WW/APRES,WIND
            COMMON /PTA/AMPAT,PAT
            COMMON /MET/ATEMP,EPRES,CLOUD,ESAT
            COMMON /N/ITPOS(20)
            COMMON /TAR/PD(101),P1(101),P2(101)
            COMMON /TAB/RD(101),R1(101),R2(101),R3(101)
            DIMENSION Z(60),Y(60)
            DIMENSION NTPL(60)
C

            DO 700 J = 1,NTEMPL
```

```
         M = J
         IF(ITPOS(J).EQ.I)GOTO 701
         AL = 0.0
  700    CONTINUE
         GOTO 702
  701    DO 720 K = 1,NTPL(M)
         Z(K) = TEMPT(1,K)
         Y(K) = TEMPL(I,K)
  720    CONTINUE
         CALL INTERP(Z,Y,T,AL,NTPL(M),1)
  702    TEMPL1 = AL
         CALL SOLRAD(T,I)
         ATEMP1 = ATEMP + (AMPAT*SIN(6.284*(T − SUNR)/PAT))
         CALL ATRAD(I,ATEMP1)
   12    PD(I) = ((P1(I) + P2(I))*DT)/D(I)
         ZZZ = (TEMPL1/(A(I)*DX))*DT
         PD(I) = PD(I)/1000.0 + ZZZ
         RETURN
         END
C
         SUBROUTINE SOLRAD(T,I)
         COMMON /MAX/CA,ALMAX,CB,QS
         COMMON /TIM/SUNR,DAYLEN
         COMMON /TAR/PD(101),P1(101),P2(101)
         COMMON /TAR/RD(101),R1(101),R2(101),R3(101)
         IF(T.LE.SUNR) GOTO 20
         IF(T.GE.SUNR + DAYLEN) GOTO 20
         AL = ALMAX*SIN((3.142/DAYLEN)*(T − SUNR))
         IF (AL.LT.0.0) GOTO 9
         GOTO 7
    9    AL = 0.1
    7    R = CA*AL**CB
         R = 0.2
         IF(R.GT.1.0) GOTO 20
         P1(I) = (1 − R)*QS*0.113
         RETURN
   20    P1(I) = 0.0
         RETURN
         END
C
         SUBROUTINE ATRAD(I,ATEMP1)
         COMMON /MET/ATEMP,EPRES,CLOUD,ESAT
         COMMON /TAR/PD(101),P1(101),P2(101)
         COMMON /TAB/RD(101),R1(101),R2(101),R3(101)
C
```

```
      TA = ATEMP1 + 9.0/5.0 + 32.0
      P2(I) = 2.05*(10.0**( − 8))*(1.0 + 0.17*CLOUD**2)*((TA + 460.0)**4)
      P2(I) = P2(I)*(1.0 + 0.149*EPRES**( − 2))
      P2(I) = P2(I)*0.113
      RETURN
      END
C
C

      SURBOUTINE SINKT(T,I,ATEMP1)
      COMMON /VEL/U(101),Q(101),A(101),D(101)
      COMMON /MESH/DX,DT,UMAX
      COMMON /MET/ATEMP,EPRES,CLOUD,ESAT
      COMMON /WW/APRES,WIND
      COMMON /WAT/WTEMP(101)
      COMMON /TAR/PD(101),P1(101),P2(101)
      COMMON /TAB/RD(101),R1(101),R2(101),R2(101)
C
      CALL BACRAD(I)
      CALL EVAP(I)
      CALL CONVEC(I,ATEMP1)
C
      RD(I) = ((R1(I) + R2(I)R3(I))*DT)/D(I)
      RD(I) = RD(I)/1000.0
      RETURN
      END
C
      SUBROUTINE BACRAD(I)
      COMMON /WAT/WTEMP(101)
      COMMON /TAR/PD(101),P1(101),P2(101)
      COMMON /TAB/RD(101),R1(101),R2(101),R3(101)
C
      TK = WTEMP(I) + 273.0
      R1(I) = 0.97*1.357*(10.0**( − 8))*(TK**4)*3600.0/1000.0
      RETURN
      END
C
      SUBROUTINE EVAP(I)
      COMMON /MET/ATEMP,EPRES,COULD,ESAT
      COMMON /WW/APRES,WIND
      COMMON /WAT/WTEMP(101)
      COMMON /TAR/PD(101),P1(101),P2(101)
      COMMON /TAB/RD(101),R1(101),R2(101),R3(101)
C
      EV = 0.000000008*WIND*(ESAT − EPRES)
      LW = 597.0 − 0.57*WTEMP(I)
```

```
          R2(I) = 1000.0*LW*EV
          R2(I) = R2(I)*3600.0
          RETURN
          END
C
          SUBROUTINE CONVEC(I,ATEMP1)
          COMMON /MET/ATEMP,EPRES,CLOUD,ESAT
          COMMON /WW/APRES,WIND
          COMMON /WAT/WTEMP(101)
          COMMON /TAR/PD(101),P1(101),P2(101)
          COMMON /TAB/RD(101),R1(101),R2(101),R3(101)
          DIMENSION TW(101)
C
```

Chapter 8

Models of Water Quality in Estuaries

A. JAMES AND D. J. ELLIOTT

8.1 INTRODUCTION

Estuaries are more complex environments than rivers, both hydraulically and biologically. Hydraulically there are four main complications:

(i) Two-directional flow (upstream and downstream) in the estuary resulting from the interaction between freshwater flow and tidal movements. This also gives rise to frequent periods of slack water.

(ii) Density differences between the freshwater (and effluents) entering an estuary and the saline water entering from the sea. This produces a highly variable degree of mixing and in cases where the mixing is poor leads to a gravitational circulation pattern.

(iii) Rivers often tend to widen out as they enter estuaries and the resultant water body is often subject to a horizontal circulation pattern due to Coriolis effects or wind-driven forces.

(iv) Near the mouth, wave action becomes important in including additional currents and in initiating or enhancing sediment transport. Waves rarely transport sediment directly except in the surf zone. But in deeper water their velocities keep sediment in suspension so that it can be transported by weak currents.

Biologically, estuaries are a distinct part of the river system. They represent a transition zone between freshwater and saline habitats, which is unfavourable to the majority of aquatic plants, animals and microorganisms. The central section of estuaries with transitional and constantly varying salinities is especially unfavourable and is often inhabited by a strictly limited flora and fauna.

An Introduction to Water Quality Modelling. 2nd Edition. Edited by A. James

From a modelling point of view, it is very important to identify the populations at risk so that appropriate environmental standards can be applied. Some standards commonly adopted for estuaries are as follows:

(a) no nuisance standard—no visible slick and dissolved oxygen (DO) $\not< 1$ mg/l;
(b) coarse fish standard—limit on toxins and DO $\not< 3$ mg/l;
(c) migratory fish standard—limit on toxins and DO $\not< 5$ mg/l.

The common feature of all these standards is the concentration of dissolved oxygen. Estuary models, like river models, therefore tend to be concerned mainly with the BOD/DO relationships. One significant difference is that due to the greater retention times in estuaries (compared with rivers) there is time for complete oxidation of the organic matter. The concept of 5-day BOD is therefore replaced by 20-day UOD.

8.2 ESTUARINE HYDRAULICS

Water movement in estuaries is the complex resultant of freshwater flow and tidal oscillation with additional momentum from the wind. In addition there are horizontal and vertical circulations brought about by Coriolis effects and density differences.

The movement of water in estuaries may be represented by a tidal excursion diagram shown in Figure 8.1. If there were no density differences in the estuary then the net resultant over a tidal cycle would be a downstream movement from AA to A'A'. Starting from high water, the water would move downstream under the combined effects of freshwater advection and tidal movement downstream and then on the flood tide move back up to A'A'.

However, in many estuaries there are sufficient density differences vertically to cause the surface water to respond more significantly to the freshwater advection and the lower layers to respond more to the tidal inflow. The resultant movement over a tidal cycle is the line BB.

Figure 8.1 also shows that variation in freshwater does not fundamentally alter this movement. High freshwater flows simply increase the size of the tidal excursion in the upper layers and still allow an upstream movement of water near the bed.

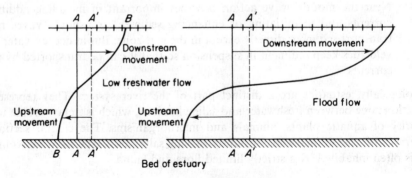

Figure 8.1. Effect of freshwater flow on tidal excursion

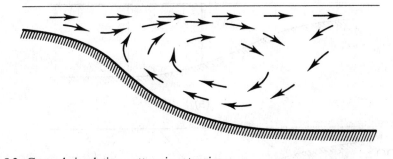

Figure 8.2. General circulation pattern in estuaries

The overall mixing pattern in estuaries may therefore be represented by a gravitational circulation as shown in Figure 8.2.

Not all hydraulic effects are significant in all estuaries and it is important, in choosing the appropriate modelling approach, to be aware of the type of estuary. Various classifications exist but the simplest uses the D/T ratio in which:

D = volume of freshwater entering an estuary during a tidal cycle;
T = volume of water exchanged at the seaward end (i.e. difference between high water and low water volumes).

When the D/T ratio is low (i.e. < 0.1) the estuary tends to be well mixed vertically and so the isohalines tend to be vertical (as shown in Figure 8.3a). When the freshwater discharge is more important (i.e. $D/T > 0.1$ and < 1.0) then the degree of mixing tends to be reduced and the isohalines are no longer vertical, see Figure 8.3b. In estuaries which are dominated by the freshwater flow (i.e. $D/T > 1.0$) mixing between fresh and saline water is extremely limited. Freshwater tends to flow over the top of a wedge of seawater, giving isohalines which are almost horizontal (see Figure 8.3c).

Topography also plays an important part in determining the mixing pattern and more complex classifications take this into account. For example, wide shallow estuaries tend to be well mixed vertically but have significant lateral differences due to streaming. Deep narrow estuaries are often well mixed laterally but have fjord-like salinity stratification.

8.3 ESTUARINE MODELS

Various techniques have been used with reasonable success in the modelling of estuarine pollution, these are as follows:

(a) Black-box—relating the concentration of the desired water quality variable, such as DO, to some easily measured environmental variables like freshwater flows.
(b) Box or segment models—representing the estuary by a series of stirred tanks.
(c) Moving segments models—representing the estuary by a series of moving boxes, each of which behaves like a stirred tank.

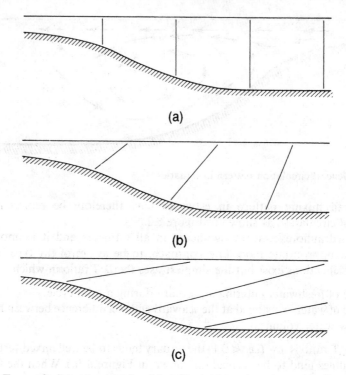

Figure 8.3. Types of mixing in estuaries; (a) well-mixed estuary; (b) partly mixed estuary; (c) stratified estuary

   (d)   Finite difference or finite element models—water quality changes are represented dynamically by following changes as the pollutants advect and disperse in fine mesh.

   Examples of all these four types of models are discussed below. As with river models it is generally assumed (except in two- or three-dimensional models) that the pollutants mix rapidly with the surrounding water so that the resultant is neutrally buoyant and has negligible effect on the existing flow pattern.

   In the choice of model it is important to consider the choice of time-scale, since there are broadly 3 categories. Firstly there are steady-state models which assume that the longitudinal distribution of any water quality parameter is identical at corresponding points and times, over successive tidal cycles. These models are often 1 dimensional, using tidally averaged parameters to depict the mid-tide motion averages to zero over a tidal cycle. This type of model is generally applied to freshwater zones of estuaries or to estuaries which are vertically well mixed.

   The second model category is that in which parameters are allowed to vary from one tidal cycle to the next. Conditions within the tidal cycle are not reproduced and advective movement is attributed solely to freshwater flow. These models are used in situations where intermittent injection of pollutants occur or where continuous injections vary slowly with a time scale longer than a tidal cycle.

The third category contains general time varying models which reproduce conditions within a tidal cycle. Advection is represented as the instantaneous resultant of tidal and freshwater flows. Such models are used in estuaries which are subject to significantly varying discharges of pollutants and freshwater flows.

There is one further consideration in estuarine models which is the way in which the hydrodynamics are incorporated into the model. Three different levels of complexity may be involved.

(a) In the simplest form of representation the water velocities are specified along with the depths. The coefficient of dispersion is either calculated from the velocity or it is specified.

(b) Models in which changes in water level are specified from which depths, cross-sectional areas and velocities can be calculated. Dispersion may be calculated or specified.

(c) The hydrodynamic representation in the most complex estuary models is based upon the equations for the conversion of mass and momentum. For a two-dimensional depth-averaged model these may be stated as follows:

(i) Conservation of mass (for water)

$$\frac{\partial n}{\partial t} + \frac{\partial(dU)}{\partial x} + \frac{\partial(dV)}{\partial y} - S = 0$$

(ii) Conservation of mass (for solutes)

$$\frac{\partial(d\bar{P})}{\partial t} + \frac{\partial(dU\bar{P})}{\partial x} + \frac{\partial(dV\bar{P})}{\partial y} - \frac{\partial\left(dD_x\left(\frac{\partial\bar{p}}{\partial x}\right)\right)}{\partial x} - \frac{\partial\left(dD_y\left(\frac{\partial\bar{p}}{\partial y}\right)\right)}{\partial y}$$

$$+ Kd\bar{P} - \bar{S} = 0$$

(iii) Conservation of momentum

$$\frac{\partial U}{\partial t} + \frac{U\partial V}{\partial x} + \frac{V\partial U}{\partial y} - fV$$

$$+ g\frac{\partial n}{\partial x} - M\left(\frac{\partial^2 U}{\partial x^2} + \frac{\partial^2 U}{\partial y^2}\right) + g\frac{d\partial p}{2p\partial x} + g\frac{U(U^2 + V^2)^{1/2}}{C^2 d} - \frac{T_{sx}}{pd} = 0$$

$$\frac{\partial V}{\partial t} + \frac{U\partial V}{\partial x} + \frac{V\partial U}{\partial y} - fU$$

$$+ g\frac{\partial n}{\partial y} - M\left(\frac{\partial^2 V}{\partial x^2} + \frac{\partial^2 V}{\partial y^2}\right) + g\frac{d\partial p}{2p\partial y} + g\frac{V(U^2 + V^2)^{1/2}}{C^2 d} - \frac{T_{sy}}{pd} = 0$$

where

U and V are depth averaged velocities in the X and Y directions;
f is the coriolis coefficient;
n is the height of the water surface above the reference plane;

h is the level of the solid boundary below the reference plane;
$d = h + n$;
M = coefficient of horizontal diffusion;
p = density of water;
C = Chezy friction coefficient;
T_{sx} and T_{sy} = shear stress components in x and y directions;
t = time;
g = gravitation coefficient;
k = decay coefficient;
$\overline{S}$ = sources and sinks averaged over depth;
$\overline{P}$ = depth averaged concentration.

Techniques for solving these equations have generally been finite difference using an implicit method. Implicit methods are preferable because of their ability to operate at high Courant numbers, avoiding the limitation imposed on explicit methods that

$$C_T = \Delta t \left(gd \left(\frac{1}{\Delta x^2} + \frac{1}{\Delta y^2} \right) \right)^{1/2} < 1$$

The usual practice is to divide the time step into two halves and to solve for the x direction in one half of the time step whilst keeping V constant and then to solve for the y direction in the other half time step while keeping U constant. This technique is known as the ADI (alternating direction implicit) method.

8.3.1 Black-box Models

An example of this category is the DO model formulated for the Clyde estuary (Mackay and Fleming 1969). As shown in Figure 8.4 the DO pattern in the estuary was represented by regression equations using data collected at nine points along the estuary. The DO at each station was related to three environmental variables which were

(a) temperature;
(b) freshwater flow;
(c) tidal amplitude.

In the data analysis a linear relationship was established between DO and temperature and DO and tidal amplitude. The relationship for flow had to be linearized by using logarithms. In evaluating this type of model there are three important considerations. Firstly the numerical values of the coefficients in the regression equations do not have any significance in judging the relative importance of the environmental variables.

Secondly the model may be used to examine the effect on DO of any combination of flow, temperature and tidal amplitude within the range of observed values. The model cannot be used to extrapolate beyond these ranges.

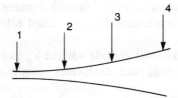

$$C_1 = a + b_1 x_1 + b_2 x_2 + b_3 x_3$$

Figure 8.4. Clyde estuary model

Thirdly the great merit of this type of model is the minimal data requirement but this also has its drawbacks. Since there was no significant change in the BOD loading during the period of water quality surveys, the BOD load does not appear as an environmental variable. The model therefore cannot be used to predict the effect of any future increase or reduction.

8.3.2 Segment Models

In this type of model the complex hydrodynamics of estuaries with the combination of advection and dispersion is reduced to a single mixing operation. It is assumed that the estuary is divided into a number of segments (as shown in Figure 8.5). Each segment is assumed to be uniformly well mixed so that the concentration in any segment may be represented by a single statistic.

In this way the estuary can be visualized as consisting of a series of continuously stirred tanks. Several models of this type have been developed but only two will be described here. The first is a simple model which can be used without the aid of digital computers and the second is a model which was used as a basis for the Thames study.

In the first of these models, the estuary is divided into a number of segments of length equal to half the mean tidal excursion. The effluent is assumed to be instantly mixed in the outfall segment and have the same characteristics as the freshwater flow. Advection and longitudinal mixing carry the effluent to adjacent segments downstream while upstream segments are affected only by longitudinal mixing. It is

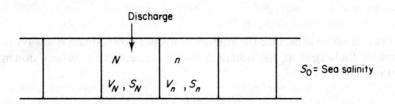

Figure 8.5. A segmented-estuary model

necessary to calculate the retention period of each segment, which is equal to that of the freshwater flow on the assumption that the added effluent does not significantly affect the mixing patterns.

Consider any segment n of total midtide volume V_n and salinity S_n.

If the seawater salinity $= S_0$ and freshwater salinity $= 0$ then the volume of seawater in the segment is

$$V_{sn} = V_n \times \frac{S_n}{S_0} \tag{8.1}$$

and the volume of the freshwater in the segment is

$$V_{fn} = V_n - V_{sn} = V_n - V_n \times \frac{S_n}{S_0} = V_n \frac{S_0 - S_n}{S_0} \tag{8.2}$$

If the freshwater flow from the river is Q during a tidal cycle, then the retention time of the freshwater in the segment is

$$R_n = \frac{V_{fn}}{Q} = \frac{V_n}{Q} \frac{(S_0 - S_n)}{S_0} \tag{8.3}$$

As the effluent is assumed to behave in the same way as the freshwater, then the effluent retention will be the same as equation (8.3).

Consider the outfall segment N into which F is the volume of conservative effluent discharged during one tidal cycle. At equilibrium, the amount of effluent entering the segment must equal the amount being advected downstream during one tidal cycle.

Taking a mass balance across the discharge segment

$$F = \frac{C_N \times V_N}{R_N} \tag{8.4}$$

Where C_N is the equilibrium concentration in segment N. Rearranging equation (8.4) gives

$$C_N = \frac{F \times R_N}{V_N}$$

$$= F \times \frac{V_N}{Q} \times \frac{(S_0 - S_N)}{S_0} \times \frac{1}{V_N}$$

$$= \frac{(S_0 - S_N)}{S_0} \times \frac{F}{Q} \tag{8.5}$$

For a conservative substance the amount conveyed downstream is always equal to the amount discharged at the outfall and the concentration in any downstream segment n is given by

$$C_n = \frac{(S_0 - S_n)}{S_0} \times \frac{F}{Q} \tag{8.6}$$

For upstream segments subject to dispersive processes only, the ratio of the upstream concentration C_m to the outfall concentration C_N will be the same as the ratio of the salinities of the two segments, giving

$$\frac{C_m}{C_N} = \frac{S_m}{S_N}$$

and

$$C_m = \frac{S_m}{S_N} \times C_N \tag{8.7}$$

Using equations (8.6) and (8.7) together with knowledge of the uniform effluent flow, uniform freshwater flow, half-tide cross-sectional average salinity data, an estimate of the concentration profile of conservative substance can be made.

This simple approach has been extended to take account of non-conservative substances. However, it is not intended to consider this model in more detail but to look at the approach used for the Thames.

Again, complete mixing within segments is assumed with steady freshwater flow and effluent flow and an equilibrium salinity distribution. Segment lengths are chosen equal to approximately one-fifth of a tidal excursion, and the continuous effluent discharge for each tide is assumed to be distributed in suitable proportions among those segments which normally receive additions during each tidal cycle.

Consider Figure 8.6, in which the estuary has been divided into segments. The effect of tidal mixing is represented by equal and opposite flows F between segments. Q, V and S are the freshwater flow half-tide volume and half-tide salinity respectively.

By taking a mass balance for salt across each segment a series of equations is obtained in which the only unknowns are the values for F:

Segment n $\qquad F_n(S_{n+1} - S_n) - (Q + Q_n)S_n = 0$

Segment $n + 1$ $\qquad F_{n+1}(S_{n+2} - S_{n+1}) + F_n(S_n - S_{n+1}) + (Q + Q_n)S_n$

$$- (Q + Q_n + Q_{n+1})S_{n+1} = 0$$

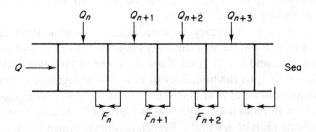

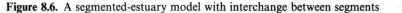

Figure 8.6. A segmented-estuary model with interchange between segments

Segment $n + 2$

$$F_{n+2}(S_{n+3} - S_{n+2}) + F_{n+1}(S_{n+1} - S_{n+2})$$
$$+ (Q + Q_n + Q_{n+1})S_{n+1}$$
$$- (Q + Q_n + Q_{n+1} + Q_{n+2})S_{n+2} = 0 \qquad (8.8)$$

Solving these equations simultaneously will give the values for F the unknown mixing coefficients. Having found the F values the model can be applied to any other dissolved or suspended substance.

If a substance C which decays exponentially is discharged into segment $n + 1$ then the rate of decay of the substance is $kC_{n+1}V_{n+1}$ where k is the decay rate constant. The mass balance equation for this segment is then

$$\frac{\partial}{\partial t}(C_{n+1}, V_{n+1}) = F_n(C_n - C_{n+1}) + F_{n+1}(C_{n+2} - C_{n+1})$$
$$+ (Q + Q_n)C_n - (Q + Q_n + Q_{n+1})C_{n+1}$$
$$- kC_{n+1}V_{n+1} + L_a = 0 \qquad (8.9)$$

where L_a is the mass added per tidal cycle. The solution to the mass balance equations for all segments yields the steady-state distribution for the concentration profile C.

This type of model would not be expected to reproduce pollutant distributions with large concentration gradients because of the step length involved. Fortunately the distribution of effluents in many estuaries exhibits shallow gradients.

Refinements to the one-dimensional mixed segment model include increasing the number of layers to describe stratified flow and also the development of a time-dependent model by advecting the boxes up- and downstream in response to tidal movements.

8.3.3 Moving Segment Models

In this approach, the zone of interest is divided into a number of segments. As the tide ebbs and flows the segments move up and down the estuary changing their shape to maintain a constant volume within each segment. If the boundaries of the segments are moving with tidal velocity U_T then the volume of the segment will remain constant with time.

Also for any segment boundary X_i the volume upstream of that boundary will remain constant with time. This fact can be used to calculate the boundary representation as shown in Figure 8.7a, b.

Figure 8.7a is a plot of estuary volume against distance from the upstream boundary of the model for three separate points in time, $t = 0$, $T/2$ and T. Where $t = 0$ is the time of low tide, and $t = T/2$ and T are the times of high tide and the next low tide respectively. Let $X_{i,0}$ be the boundary of a particular segment at low tide and let V_i be the upstream volume of that boundary. From the previous discussion V_i is to remain constant for all t and may be plotted in Figure 8.7a as a horizontal line which intersects the volume plots at a, b and c. Perpendiculars dropped to the x-axis identify the position of the boundary low tide, mean tide and high tide. If the volume–distance

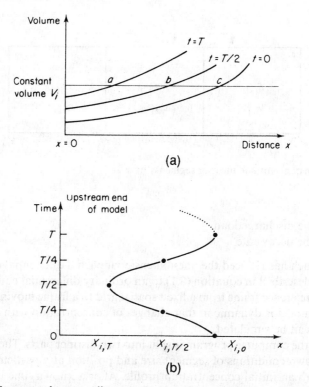

Figure 8.7. (a) Estuary volume vs. distance; (b) identifying the segment boundary

graph is plotted for intermediate times, a time–distance graph can be constructed as shown in Figure 8.7b, identifying the boundary at any time during the tide cycle. This information can be computed for each segment boundary and hence the centre of each segment can be established at any point in time.

In Figure 8.8 consider the boundary X_i between segment i and $i + 1$. The transport of material through the boundary is made up of two terms.

(1) due to freshwater flow, $Q_i C_i$, where $Q_i = A_i(U_i - U_{Ti})$
    A_i = cross-sectional area at i;
    U_i = velocity of freshwater;
    U_{Ti} = velocity of the tide;

(2) due to longitudinal dispersion, simulated by assuming equal and opposite flow between segments, $F_i(C_i - C_{i+1})$.

It is now possible to write a mass balance for segment i giving

$$V_i \frac{dC_i}{dt} = Q_{i-1} - C_{i-1} + F_{i-1}(C_{i-1} - C_i) - Q_i C_i$$

$$+ F_i(C_{i+1} - C_i) - K_i V_i C_i + L_{ai} \tag{8.10}$$

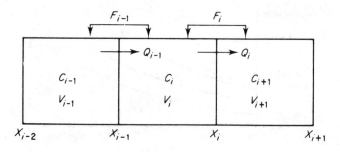

Figure 8.8. Constant-volume moving segments

where L_{ai} is the discharge load
K_i is the decay rate.

This approach has reduced the one-dimensional partial differential equation of the general form described in equation (8.1) to an ordinary differential equation (8.10) by changing the reference frame from a fixed spatial grid to a frame moving with the tidal motion. The model is dynamic in that changes of concentration with respect to time and distance can be simulated.

Essentially the computer program is split into two distinct parts. The first stage is to start with known conditions of segment size and position at a particular stage in the tidal cycle with an initial concentration profile. After a known time increment Δt, a new tide height can be calculated using empirical data on tidal behaviour. From this the new tidal volume can be calculated from which the movement of segment boundaries can be determined. Having found the new position of the segments the second stage of the computation is to solve equation (8.10) for each segment.

The Crank–Nicholson approximation may be used to numerically solve equation (8.10) for small increments of time, Δt. The time derivative dC_i/dt may be represented by a forward difference approximation over the time interval j to $j + 1$ by

$$\frac{C_i^{j+1} - C_i^j}{\Delta t}$$

The right-hand side of the equation (8.10) can be approximated by the average of values at j and $j + 1$ time steps. In other words, the difference scheme is centred in time about point $j + 1/2$.

Equation (8.10) now becomes

$$
\begin{aligned}
(V_i^{j+1}C_i^{j+1} - V_i^j C_i^j) = {} & \tfrac{1}{2}\,\Delta t(Q_{i-1}^{j+1}C_{i-1}^{j+1} - Q_i^{j+1}C_i^{j+1} \\
& + F_{i-1}^{j+1}(C_{i-1}^{j+1} - C_i^{j+1}) + F_i^{j+1}(C_{i+1}^{j+1} - C_i^{j+1}) \\
& + K_i V_i^{j+1} C_i^{j+1}) \\
& + \tfrac{1}{2}\,\Delta t(Q_{i-1}^j C_{i-1}^j - Q_i^j C_i^j + F_{i-1}^j(C_{i-1}^j - C_i^j) \\
& + F_i^j(C_{i+1}^j - C_i^j) + K_i V_i^j C_i^j)
\end{aligned}
\tag{8.11}
$$

Each equation for each segment has three unknown values of $C_{i-1}^{j+1}, C_i^{j+1}, C_{i+1}^{j+1}$ and these can be arranged in the form

$$\alpha_i\, C_{i+1}^{j+1} + \beta_i\, C_{i+1}^{j+1} + \gamma_i\, C_{i+1}^{j+1} = \delta_i(C_{i-1}^j, C_i^j, C_{i-1}^j) \tag{8.12}$$

These can be solved simultaneously using techniques described in the next section and the values obtained used as initial values for the next time increment. The solution is therefore iterative. At each increment of time the positions of the segments are calculated first and then the new concentrations in each segment.

The exchange coefficients F_i may be found using salinity data and assuming that there is a balance between the salt transported between the segments by the freshwater velocities and the exchange flows, e.g.

$$F_i(S_{i+1} - S_i) = Q_i\, S_i$$

$$F_i = \frac{Q_i S_i}{(S_{i+1} - S_i)} \tag{8.13}$$

On substituting equation (8.13) into equation (8.11) and using the known salinity profile for initial conditions, the model may be run for a series of time steps to check that the salinity in each box remains constant in time.

One advantage of this type of model is that conceptually it is easy to visualize the representation of the advection and dispersion mechanisms. A second significant advantage is that this formulation reduces the tendency for numerical dispersion to occur compared with the fixed grid finite difference approach.

Models of this type have been successfully used on the Thames and Humber estuaries.

8.3.4 Finite Difference Estuary Models

Several models of one and two dimensions have been formulated based on numerical solution of the convective diffusion equation using finite difference techniques. The following example describes the solution to the one-dimensional equation using an implicit difference approximation.

The general form of the convective diffusion may be used to describe the estuary situation:

$$\frac{\partial AC}{\partial t} = \frac{-\partial AUC}{\partial x} + \frac{\partial}{\partial x}\left(EA\frac{\partial C}{\partial x}\right) - KAC + L_a \tag{8.14}$$

The cross-sectional area A, effective longitudinal dispersion coefficient E and advective velocity U are functions of time t and distance x. Figure 8.9 shows the distance time plane covered by a mesh of sides $\delta x = L$ and $\delta t = K$. The implicit solution steps through time by increments of K and at each new point in time evaluates the values of C for all mesh points along the x direction.

For each point on the $j + 1$ row the derivatives in equation (8.10) are estimated using the values of the six points shown in Figure 8.9. Thus for n distance mesh points $n - 2$, simultaneous equations are developed, each of which has three unknowns

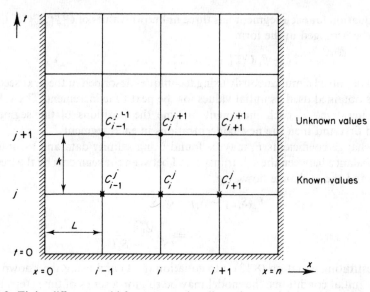

Figure 8.9. Finite difference grid for an estuarine dispersion model

except for the equations at each end. They only have two unknowns at the $j + 1$ level because the boundaries of the mesh are chosen to produce known values of concentration at all time intervals. These known boundary values are essential for solving the simultaneous equations as shown below. The implicit solution is started by assuming an initial concentration distribution at $t = 0$. This particular formulation produces a dynamic solution which varies with time and the model is run until the particular dynamic effect of a transient load is represented or until a pseudo-steady state is reached for uniform loads at which concentrations do not vary between equivalent points in the tidal cycle.

A typical implicit approximation method (Crank–Nicholson) approximates derivatives in the x direction, i.e. $\partial/\partial x$ and $\partial^2/\partial x^2$ by the mean of its finite difference representation of the $j + 1$ and the j rows.

In calculating C_i^{j+1} the time derivative $\partial/\partial t$ used to advance the computation is computed from a weighted average of the derivatives at the distance points $i - 1$, i and $i + 1$. The weighting used here is that giving a stable solution with minimal numerical dispersion, which is insignificant if the value of E is reasonably large.

Each term in equation (8.14) is represented separately below by its difference approximation.

(a) *Time derivative*

$$\frac{\partial AC}{\partial t} \simeq \frac{1}{6}\left(\frac{A_{i-1}^{j+1}C_{i-1}^{j+1} - A_{i-1}^{j}C_{i-1}^{j}}{k}\right) + \frac{2}{3}\left(\frac{A_i^{j+1}C_i^{j+1} - A_i^{j}C_i^{j}}{k}\right)$$

$$+ \frac{1}{6}\left(\frac{A_{i+1}^{j+1}C_{i+1}^{j+1} - A_{i-1}^{j}C_{i-1}^{j}}{k}\right) \tag{8.15}$$

here the differences at $i - 1$, i, $i + 1$ are weighted $\frac{1}{6}$, $\frac{2}{3}$, $\frac{1}{6}$ respectively.

(b) *Advective term*

$$\frac{\partial U A C}{\partial x} \simeq \frac{1}{2}\left(\frac{U_{i+1}^{j+1}A_{i+1}^{j+1}C_{i+1}^{j+1} - U_{i-1}^{j+1}A_{i-1}^{j+1}C_{i-1}^{j+1}}{2h}\right)$$

$$+ \frac{1}{2}\left(\frac{U_{i+1}^{j}A_{i+1}^{j}C_{i+1}^{j} - U_{i-1}^{j}A_{i-1}^{j}C_{i-1}^{j}}{2h}\right) \qquad (8.16)$$

equal weight is given to j and $j + 1$ differences which both use values at mesh points $i + 1$ and $i - 1$ divided by twice the x mesh distance to approximate the derivative at mesh distance i.

(c) *Dispersive term*

$$\frac{\partial}{\partial x} E A \frac{\partial C}{\partial x} \simeq \frac{1}{h} \delta x (EA)_i \frac{1}{h} \delta x \, C_i^j \qquad (8.17)$$

where δx is called the central difference operator, i.e.

$$\delta x \, C_i^j = C_{1+\frac{1}{2}}^j - C_{i-\frac{1}{2}}^i$$

and

$$\frac{\partial C}{\partial x} \simeq \frac{\delta x \, C_{ij}}{h}$$

the derivative being evaluated at the point i, j.

Hence

$$\frac{1}{h} \delta x [(EA)_i \frac{1}{h} \delta x \, C_i] = \frac{1}{h} \delta x [(EA)_i \frac{1}{h} (C_{i+1/2} - C_{i-1/2})]$$

$$- \frac{1}{h^2} \{[(EA)_{i+1/2} \, C_{i+1} - (EA)_{i-1/2} \, C_i]$$

$$- [(EA)_{i+1/2} \, C_i - (EA)_{i-1/2} \, C_{i-1}]\} \qquad (8.18)$$

Thus the dispersive term is approximated using values of C at $i - 1$, i and $i + 1$ but E and A are evaluted at the half mesh point in the x direction (see below).

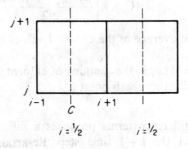

Figure 8.10. Subset of time–distance mesh

Now

$$(EA)_{i+1/2} = \frac{(EA)_{i+1} + (EA)_i}{2}$$

and

$$(EA)_{i-1/2} = \frac{(EA)_i + (EA)_{i-1}}{2}$$

Substituting in equation (8.18) gives

$$\frac{1}{h^2}\left\{\left[\left(\frac{(EA)_{i+1} + (EA)_i}{2}\right)C_{i+1} - \left(\frac{(EA)_i + (EA)_{i-1}}{2}\right)C_i\right]\right.$$
$$\left.- \left[\left(\frac{(EA)_{i+1} + (EA)_i}{2}\right)C_i - \left(\frac{(EA)_i + (EA)_{i-1}}{2}\right)C_{i-1}\right]\right\} \quad (8.19)$$

and grouping terms gives

$$\frac{1}{h^2}\left[\left(\frac{(EA)_{i+1} + (EA)_i}{2}\right)(C_{i+1} - C_i) - \left(\frac{(EA)_i + (EA)_{i-1}}{2}(C_i - C_{i-1})\right)\right]$$
$$(8.20)$$

Now the dispersive derivative is approximated in equation (8.20) by the variable coefficients E and A evaluated at the mesh points.

As mentioned earlier, the Crank–Nicholson approximation uses equally weighted differences at the j and $j + 1$ time intervals, so that

$$\frac{\partial}{\partial x} EA \frac{\partial C}{\partial x} \simeq \frac{1}{2} \cdot \frac{1}{2h^2} (\{[(EA)_{i+1}^{j+1} + (EA)_i^{j+1}](C_{i+1}^{j+1} - C_i^{j+1})\}$$
$$- \{[(EA)_i^{j+1} + (EA)_{i-1}^{j+1}](C_i^{j+1} - C_{i-1}^{j+1})\})$$
$$+ \frac{1}{2} \cdot \frac{1}{2h^2} (\{[(EA)_{i+1}^j + (EA)_i^j(C_{i+1}^j - C_i^j)\}$$
$$- \{[(EA)_i^j + (EA)_{i+1}^j](C_i^j - C_{i-1}^j)\}) \quad (8.21)$$

(d) *Reaction term*

$$kAC \simeq k\left[\frac{(AC)_i^{j+1} + (AC)_i^j}{2}\right] \quad (8.22)$$

This term uses the average of the C and A values at the j and $j + 1$ row.

(e) *Load term*
$La = \Delta LA_i$ where ΔLa_i is the amount of effluent discharged over the time interval k at the distance mesh point i.

The combination of difference terms produces a difference formula with three unknown values of C at the $j + 1$ time step. Re-arranging this expression by combining terms an equation may be produced with the three unknowns on the left-hand side and known values on the right-hand side, i.e.

$$- \alpha_i^{j+1} C_{i-1}^{j+1} \beta_i^{j+1} C_i^{j+1} - \gamma_i^{j+1} C_{i-1}^{j+1} = \delta_i^j \tag{8.23}$$

where

α and γ are functions of A, E, U at $j + 1$
β is a function of A and E at $j + 1$
δ is a function of A, E, U, C at j

Equation (8.19) is one of a system of n linear algebraic equations where $1 < i < n - 1$ and C_0 and C_n are known from boundary conditions. If

$$\alpha_i > 0 \qquad \beta_i > 0 \qquad \gamma_i > 0$$

and

$$\beta_i \geq (\alpha_i + \gamma_i)$$

then this system may be solved.

When written in matrix notation it may be seen that a tridiagonal form is produced. An efficient solution equivalent to Gaussian elimination is available to solve the system, which may be written as

$$\begin{bmatrix} \beta_1 & -\gamma_1 & & & \\ -\alpha_2 & \beta_2 & -\gamma_2 & & \\ & & \ddots & \ddots & \\ & & -\alpha_{n-2} & \beta_{n-2} & \gamma_{n-2} \\ & & & -\alpha_{n-2} & \beta_{n-2} \end{bmatrix} \begin{bmatrix} C_1 \\ C_2 \\ \vdots \\ C_{n-2} \\ C_{n-1} \end{bmatrix} = \begin{bmatrix} \delta_1 + C_0 \\ \delta_2 \\ \vdots \\ \delta_{n-2} \\ \delta_{n-1} + \gamma_{n-1} C_n \end{bmatrix} \tag{8.24}$$

Let us assume that $C_0 = 0$ and $C_n = K$.

The first equation may be written as

$$\beta_1 C_1 - \gamma_1 C_2 = \delta_1 \tag{8.25}$$

and the second equation is

$$- \alpha_2 C_1 + \beta_2 C_2 - \gamma_2 C_3 = \delta_2 \tag{8.26}$$

from equation (8.25)

$$C_1 = \frac{\delta_1 + \gamma_1 C_2}{\beta_1} = \frac{\gamma_1}{\beta_1} C_2 + \frac{\delta_1}{\beta_1} \tag{8.27}$$

substituting for C_1 in (8.26) gives

$$- \alpha_2 \left(\frac{\gamma_1 C_2}{\beta_1} + \frac{\delta_1}{\beta_1} \right) + \beta_2 C_2 - \gamma_2 C_3 = \delta_2 \tag{8.28}$$

Rearranging for C_2

$$C_2\left(\beta_2 - \frac{\alpha_2\gamma_1}{\beta_1}\right) = \delta_2 + \frac{\alpha_2\delta_1}{\beta_1} + \gamma_2 C_3$$

$$C_2 = \frac{\delta_2 + \alpha_2(\delta_1/\beta_1)}{\beta_2 - \alpha_2(\gamma_1/\beta_1)} + \frac{\gamma_2}{\beta_2 - \alpha_2(\gamma_1/\beta_1)} C_3 \tag{8.29}$$

C_2 may now be substituted in the next equation and C_3 expressed in terms of C_4.
On inspection of equation (8.27) this may be rewritten as

$$C_1 = a_1 C_2 + b_1 \tag{8.30}$$

where

$$\alpha_1 = \frac{\gamma_1}{\beta_1} \quad \text{and} \quad b_1 = \frac{\delta_1}{\beta_1} \tag{8.31}$$

In general, an expression for C_i may be written in terms of C_{i+1} in the form

$$C_i = a_i C_{i+1} + b_i \tag{8.32}$$

where

$$a_i = \frac{\gamma_i}{\beta_i - \alpha_i a_{i-1}}, \quad b_i = \frac{\delta_i + \alpha_i b_{i-1}}{\beta_i - \alpha_i a_{i-1}}$$

with

$$C_0 = 0 \quad \text{then} \quad a_0 = 0 \quad \text{and} \quad b_0 = 0 \tag{8.33}$$

hence

$$C_0 = a_0 C_i + b_0 \tag{8.33}$$

equation (8.33) holds true for any value of C_i.

Starting with the first equation in the system, equation (8.32) can be used to successively eliminate C_i from the next equation in the system.

When the last equation is reached $C_n = K$ and

$$C_{n-1} = a_{n-1}K + b_{n-1} \tag{8.34}$$

from which C_{n-1} can be calculated using the known values of a_{n-1}, K and b_{n-1}.

All remaining C_i can now be calculated by back substitution, e.g.

$$C_{n-2} = a_{n-2}C_{n-1} + b_{n-2}$$

$$\vdots$$

$$C_1 = a_1 C_2 + b_1 \tag{8.35}$$

To avoid substantial errors with this method it is necessary that

$$0 < a_i \le 1 \quad \text{for} \quad 1 \le i \le n-1$$

Once the values of C_i^{j+1} have been calculated, they become the initial values for the next time step and the process is repeated.

As mentioned earlier, α, β and γ are functions of the cross-sectional area A, dispersion coefficient E and velocity U. It is necessary that each of these is calculated at time $j + 1$ for each i mesh point in order that the system of equations described in equation (8.23) can be solved for the concentration distribution at $i + 1$.

The velocity U_i^{j+1} may be calculated in two ways. Firstly, U_i^{j+1} can be obtained from a solution of the equations of motion which may be developed in a similar way to the convective diffusion equation by balancing the forces acting on an element of volume of water. The equation of motion itself must be solved by a numerical technique similar to that described above. Using the velocity in the continuity equation, the tidal height can be calculated and hence the cross-sectional area at time $j + 1$.

An alternative method of calculating U is to assume that the water level in the estuary is essentially horizontal and that it varies sinusoidally according to the relationship

$$h = h_{mean} - h_{amp} \cos \omega t \qquad (8.36)$$

where

h_{mean} is the mean tide height
h_{amp} is the tidal amplitude
ω is $2\pi/12.4$ (assuming a tidal period of 12.4 hours)
t is time taken as zero at low tide

For a given time interval k the new tide height of $j + 1$ can be calculated. Knowing the hydrographic data of the estuary it is possible to calculate the change in volume upstream of any mesh point i during the time interval k. This change in volume is equal to the flow through the cross-sectional area at i hence the velocity U_i^{j+1} can be calculated.

The one-dimensional dispersion coefficient for a homogeneous estuary zone may be calculated from an expression such as equation (8.37).

$$E = 63nU/R \qquad (8.37)$$

where

n is Manning's roughness coefficient
R is the hydraulic radius (m)
E is in m^2/s
U is in m/s

Equation (8.37) may be extended to zones which have longitudinal salinity gradients because of the averaging process to reduce the convective diffusion equation from three to one dimensions. It is therefore unique to the particular estuary under consideration. In practice, values included in the expression describing the dispersion coefficient are obtained from field observation using fluorescent or radioactive tracers. Even these may only give an order of magnitude result and a fine tuning exercise is usually carried out by a curve-fitting process using the salinity profile. In other words, salinity data are collected from the estuary and the high- and low-tide salinity profiles

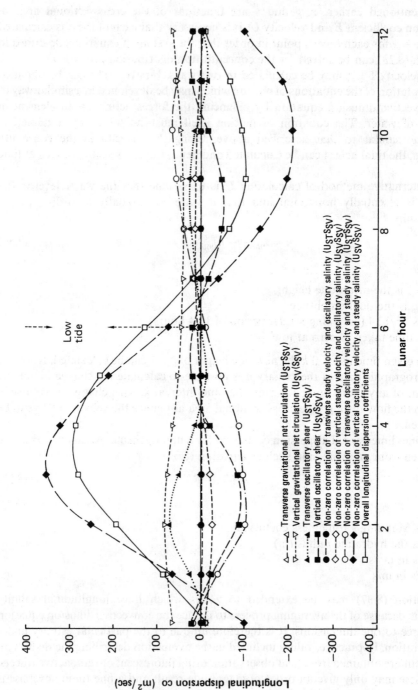

Figure 8.11. Tidal variations of decomposed longitudinal dispersion coefficients from observed velocity and salinity at Hebburn (Station 3)

are predicted over time using the finite difference scheme above with no decay or load terms. If the salinity profiles remain constant over a series of tidal cycles it may be assumed that the hydrodynamics of the estuary are correctly described. If the profiles alter significantly from the starting profile then adjustments are made to the dispersion coefficient.

Having successfully predicted the salinity distribution the model can be used to fit the DO sag curve and the UOD profile to existing data using the known existing discharge loads. The biochemical rate coefficients can be adjusted at this state. When existing conditions can be satisfactorily represented, predictions can be made of the UOD/DO profiles for different loading conditions.

8.4 DATA REQUIREMENTS

Estuaries are a particularly difficult environment for data collection. The validation and calibration of numerical models generally require collection of the following data simultaneously at several sections along the estuary, at several depths and at several time intervals over a complete tidal cycle.

Data needed are:

(a) water levels and water depths;
(b) velocities and directions;
(c) salinities and temperatures;
(d) water quality parameters like DO, BOD;
(e) loads of pollutants entering the estuary;
(f) freshwater flows from any significant tributaries;
(g) estimations of coefficients of dispersion, re-aeration, etc.

The number of cross-sections needed is very difficult to define, as is the degree of detail at each cross-section. Where lateral or vertical mixing is good, centre-line data or depth averaged data may be sufficient. But even in these cases the logistics of supplying several boats with crews and instruments make estuarine surveys extremely expensive.

Also the quality of data obtained has in the past been too unreliable to properly validate estuarine models. Table 8.1 shows the results of a sensitivity analysis which highlights this difficulty. Fortunately improvements in instrumentation through position-fixing, radar and ultrasonic measurements of velocity, recording from specific ions, electrodes, etc., are revolutionizing the quality of data available from estuaries.

The data are usually collected in the summer with low freshwater flow and maximum temperature, which should theoretically represent the 'worst' condition in the estuary. If water quality objectives can be satisfied at this time of year, then they can usually be achieved throughout the year.

A typical survey lasts for about a week, usually covering a neap or a spring tide period, during which the freshwater flow does not vary significantly. Several survey stations are sited along the zone of interest and samples are taken at 30- or 60-minute

Table 8.1. Sensitivity analysis of estuarine coefficients

Coefficient	Effect of 50% variation in coefficient on DO	Likely range of error in field measurement
Dispersion	± 50 – 100%	± 50–100%
Reaeration	± 20–80%	not normally measured
Sediment oxygen demand	± 50–100%	± 20–80% depends on variability of sediment
BOD	± 20–50%	± 20%
Photosynthesis	± 10–80%	± 50%

intervals over the period of a complete tidal cycle. From this data a series of snapshots can be obtained in the form of longitudinal profiles.

Alternatively in a short estuary it may be possible by high-speed boat or helicopter to make a water quality survey in the estuary over a fairly short period (e.g. an hour). This can be used to give similar 'instantaneous' snapshots, but it cannot be used for collecting hydraulic data.

This describes the movement of materials such as pollutants, due to advection. Pollutants also move, in response to concentration gradients, by eddy diffusion in a similar manner to the dispersion in rivers. The convective–diffusion equation may therefore be applied in estuaries, particularly to the one-dimensional representation, as shown in Figure 8.1. Upstream advection is often a significant phenomenon, so there is a danger in using a one-dimensional representation since the upstream movement is represented as longitudinal dispersion. The fallacy of this approach is illustrated in Figure 8.11, which shows that the coefficient of longitudinal dispersion regularly becomes negative during the flood tide.

REFERENCE

Mackay, D. W. and Fleming (1969) Correlation of dissolved oxygen levels, freshwater flows and temperature in a polluted estuary, *Water Research*, **3**, 121–4.

APPENDIX A

1 MIXED SEGMENT MODEL

The following listing gives a program for the segment model described in Section 8. It is formulated to calculated the distribution of BOD or other pollutant which undergoes first-order decay. But it may readily be modified to describe other types of pollutants.

The variable names have the following meanings

V = volume
S = salinity
L = BOD
Q = flow of freshwater
E = flow of effluent
D = discharge reference number
R = retention time
K = decay coefficient
N = segment reference number
SI = salinity of seawater

```
'segment model
'
DIM v(10), s(10), b(10), r(10)
FOR w = 1 TO 10
      READ v(w), s(w)
NEXT w
'
k = .05
'
INPUT S1, q
FOR d = 1 TO 3
    INPUT n, e
    t = n
120 r(t) = r6 + (v(t)/q)
    r6 = r(t)
    1(t) = (e/2*q)*(1 + 10^( - k*r6))*((s1 - s(t))/s1)
    PRINT "1(t) = ", 1(t)
    1(6) = 1(t)
    b(t) = b(t) + 1(t)
    s(6) = s(t)
    FOR t = (n - 1) TO 1 STEP - 1
        r(t) = r(5) + (v(t)/q)
        r(5) = r(5)
        1(t) = 16*(s(t)/s(t + 1))*(10^( - k*r5))
        PRINT"1(t) = ", 1(t), "t = ", t
        b(t) = b(t) + 1(t)
        u5 = u5 + v(t)*1(t)
    NEXT t
    e1 = e - u5
    IF ((e - e1) > .5) THEN 120
    '
```

```
      e = el
      t = n
      FOR t = n TO 10
         r(7) = r(t)
         1(t) = (e/2*q)*(10^( − k* r7))*(1 + 10^( − k*r7))*((s1 − s(t))/s1)
         PRINT "1(t) = ", 1(t), "t = ", t
         b(n) = b(n) + 1(t)
      NEXT t
   NEXT d
   PRINT "BOD concentrations"
   '
   FOR d = 1 TO 10
      PRINT b(d)
   NEXT d
   DATA .5,.6,.5,1.6,.5,2.8,1,3.9,1,7.6
   DATA 1.2, 10,1.4,13,1.5,18,3,22,22,28
   END
```

2   FINITE DIFFERENCE MODEL

In this model an estuary consists of a series of computational elements. For each element the hydrological mass balance can be written in terms of

- (a)   flows into and out of the upstream face;
- (b)   flows into and out of the downstream face;
- (c)   lateral inflows or abstractions.

The overall mass balance on each element is calculated from the change in water level, which determines the cumulative volume upstream of each mode at each time step.
   Similarly a material balance for any pollutant can be constructed from

- (a)   advective movements into and out of the segment;
- (b)   dispersive transfer into and out of the segment;
- (c)   rate of formations and decay;
- (d)   additions and removals.

The general mass balance may be represented by the convective diffusion equation

$$\frac{\partial AC}{\partial t} = -\frac{\partial AUC}{\partial x} + \frac{\partial}{\partial x}\left[EA\frac{\partial C}{\partial x}\right] \pm KAC \pm La$$

Where A = cross-sectional area
   U = velocity
   E = longitudinal dispersion coefficient
   x = distance downstream
   C = concentration

K = rate coefficient
La = source and sink

The area, velocity and dispersion coefficient are all functions of time and distance. Tributary and discharges are represented as loads of BOD, DO, ammonia and coliforms. Freshwater flow and tidal flow are represented dynamically in the model as are the loads and the boundary conditions.

Although not a pollutant, salinity has been included in the model since as a conservative tracer it assists with calibration and verification.

```
C
C
C       *******************************************************
C       *************FINITE DIFFERENCE MODEL*************
C       **********        MARCH 1985
C
C       *************MODIFIED FEBRUARY 1989**************
C
C       ****LAST MODIFICATION FEB/MAR/APR/MAY 1990****
C
C
C       **************data input revised june 1990***************
C
C       *****COLIFORM SUBROUTINE ADDED SEPT 1990*****
C
C       **********FINAL MODIFICATIONS MAY 91***********
C       ****************MAIN PROGRAMME*****************
C
$DEBUG
            COMMON /PHOT/BIO(101),PMAX,DAYLEN,TS
            COMMON /DOX/CS,AK2,PDO(101),ADO(101)
            COMMON /DOX1/D(101),CDO(101),DOL(101,50),
                      DOTT(101,50)
            COMMON /DOX2/TTDO(20,50),DL(20,50)
            COMMON /VEL/U(101),Q(101),A(101),W(101)
            COMMON /COEF/AK1,AK4,AK5,AK6
            COMMON /BOX1/CBOD(101),BODL(101,50),BODT(101,50)
            COMMON /RESP/BRES(101)
            COMMON /MESH/DX,DT
            COMMON /AMON/ACONC(101),AML(101,50),AMT(101,50)
            COMMON /WDTH/WDIST(20),WIDTH(20),DSDIST(20),
                      DIS(20)
            COMMON /FD/FLDIST(20),FL(20)
            COMMON /VD/VDIST(20),VEL(20)
            COMMON /DDT/DDIST(20),DEPTH(20)
            COMMON /DBT/DISTBL(20),BL(20,50),TTBOD(20,50)
            COMMON /DIL/DISTD(20), DISTDL(20),DISTCL(20)
```

```
        COMMON /DAL/DISTAL(20),AL(20,50),TTAM(20,50)
        COMMON /DIA/DISTA(20),ACC(20)
        COMMON /DIB/DISTB(20),BOD(20)
        COMMON /DIO/DISTDO(20),DO(20)
        COMMON /DBMS/DISTBM(20),BIOMS(20)
        COMMON /DBEN/DISTBR(20),BENR(20)
        COMMON /DTP/DOT(20),DTEMP(20)
        COMMON /NN/NWIDTH,NFL,NVEL,NBODL,NDOL,
                 NDIS,NAL,NACONC
        COMMON /NNN/NBOD,NDO,NBIOMS,NBENR
        COMMON /IFD/IFLOW,DISP
        COMMON /TTT/TSTART,TSTOP,TPRINT
        COMMON /TT/TLEN/TEMP
        COMMON /DF/DFACT(20),NFACT,DF(101),DFDIS(101)
        COMMON /VIC/VC(101),V1C(101),DNEW(101)
        COMMON /U/UU(101)
        COMMON /SS/DISAL(20),SAL(20),S(101),DUMY(101)
        COMMON /ILPOS/IBPOS(20), IAPOS(20),IDPOS(20),
                 ICPOS(20)
        COMMON /COLI/CCONC(101),CLL(101,50),CLT(101,50)
        COMMON /NLOD/NBLD(20),NDL(20),NALD(20),NCLD(20)
        COMMON /CHL/TTCL(20,50),CL(20,50)
C
C
        DIMENSION C(101),F(101),X(101)
        DIMENSION SIG(101),CR(101),E(101),DOLD(101)
        DIMENSION PD(101),RD(101),PB(101),RB(101),PA(101),
          RA(101)
        DIMENSIONPC(101),RC(101)
        DIMENSION DISTC(20),CCC(20)
        DIMENSION TX(20),TY(20)
C
        CHARACTER*64 HEAD
        CHARACTER*9 FNAME
C
        OPEN(4,FILE = 'USER')
        WRITE(*,'(A)')'SPECIFY DATAFILE'
        READ(*,'(A)') FNAME
        OPEN(5, FILE = FNAME)
        OPEN(6,FILE = 'OUTPUT.EST')
C
C
  901   FORMAT (A64)
  919   FORMAT (3XA64)
  902   FORMAT (A9)
C
```

```
C
C
                READ(5,901)HEAD
                READ(5,602)TLEN,DX,DT
                NMESH = INT(TLEN/DX) + 1
                WRITE(6,902) 'Graphfile'
                WRITE(6,631)NMESH
                X(1) = 0.0
                DO 540 I = 2, NMESH
                X(I) = X(I − 1) + DX
   540          CONTINUE
                READ(5,602) TSTART,TSTOP,TPRNT

                READ(5,901)HEAD
                READ(5,602) AMAN,DISP,DAMPL
                READ(5,602) ULEN,V1M,AK1
                READ(5,602) AK5,AK6,AK9
                READ(5,602) TEMP,PMAX,DAYLEN
                READ(5,602) TS,THW
                READ(5,600) MTIDE
C
                DO 1209 I = 1, MTIDE
                READ(5,602)TX(I),TY(I)
  1209          CONTINUE

C               SET UPSTREAM BOUNDARY CONDITION
C
                READ(5,901)HEAD
                WRITE(4,919)HEAD
                READ(5,901)HEAD
                WRITE(4,919)HEAD
                READ(5,602) UBOD,AMPB,PERB
                READ(5,602) UDO,AMPD,PERD
                READ(5,602) UAM,AMPA,PERA
                READ(5,602) UAS,AMPS,PERS
                READ(5,602) UC,AMPC,PERC
                WRITE(4,602) UBOD,AMPB,PERB
                WRITE(4,602) UDO,AMPD,PERD
                WRITE(4,602) UAM,AMPA,PERA
                WRITE(4,602)UAS,AMPS,PERS
                WRITE(4,602) UC,AMPC,PERC
                CBOD(1) = UBOD
                CDO(1) = UDO
                ACONC(1) = UAM
                S(1) = UAS
```

```
              UC = 10.0**(UC)
              CCONC(1) = UC
C
C
              SET DOWNSTREAM BOUNDARY CONDITIONS
              READ(5,901)HEAD
              WRITE(4,919)HEAD
              READ(5,602) DBOD,DAMPB,DPERB
              READ(5,602) DDO,DAMPD,DPERD
              READ(5,602) DAM,DAMPA,DPERA
              READ(5,602) DS,DAMPS,DPERS
              READ(5,602) DC,DAMPC,DPERC
              WRITE(4,602) DBOD,DAMPB,DPERB
              WRITE(4,602) DDO,DAMPD,DPERD
              WRITE(4,602) DAM,DAMPA,DPERA
              WRITE(4,602) DS,DAMPS,DPERS
              WRITE(4,602) DC,DAMPC,DPERC
              CBOD(NMESH) = DBOD
              CDO(NMESH) = DDO
              ACONC(NMESH) = DAM
              S(NMESH) = DS
              DC = 10.0**(DC)
              CCONC(NMESH) = DC
C    INITIAL CONDITIONS BOD AND DO
C
              READ(5,901)HEAD
              WRITE(4,919)HEAD
              READ(5,901)HEAD
              WRITE(4,919)HEAD
              READ(5,600)NBOD
              WRITE(4,600)NBOD
              READ(5,601)(DISTB(I),BOD(I),I = 1,NBOD)
              WRITE(4,601)(DISTB(I),BOD(I),I = 1,NBOD)
              DO 1500 I = 1,NMESH
              XI = X(I)
              CALL INTERP(DISTB,BOD,XI,CBOD(I),NBOD,1)
1500          CONTINUE
              READ(5,901)HEAD
              WRITE(4,919)HEAD
              READ(5,600)NDO
              WRITE(4,600) NDO
              READ(5,601)(DISTDO(I),DO(I),I = 1,NDO)
              WRITE(4,601)(DISTDO(I),DO(I),I = 1,NDO)
              DO 1600 I = 1,NMESH
              XI = X(I)
```

```
          CALL INTERP (DISTDO,DO,XI,CDO(I),NDO,1)
1600      CONTINUE
C
C         SET INITIAL CONDITIONS FOR AMMONIA
C
          READ(5,901)HEAD
          WRITE(4,919)HEAD
          READ(5,600) NACONC
          WRITE(4,600) NACONC
          READ(5,601)(DISTA(I),ACC(I),I = 1,NACONC)
          WRITE(4,601)(DISTA(I),ACC(I),I = 1,NACONC)
          DO 1650 I = 1,NMESH
          XI = X(I)
          CALL INTERP(DISTA,ACC,XI,ACONC(I),NACONC,1)
1650      CONTINUE
C
C
C
C         SET INITIAL CONDITIONS FOR SALINITY
C
          READ(5,901)HEAD
          WRITE(4,919)HEAD
          READ(5,600)NSAL
          WRITE(4,600)NSAL
          READ(5,601)(DISAL(I),SAL(I),I = 1,NSAL)
          WRITE(4,601)(DISAL(I),SAL(I),I = 1,NSAL)
          DO 1651 I = 1,NMESH
          DUMY(I) = 0.0
          XI = X(I)
          CALL INTERP(DISAL,SAL,XI,S(I),NSAL,1)
1651      CONTINUE
C
C         SET INITIAL CONDITIONS FOR COLIFORMS
C
          READ(5,901)HEAD
          WRITE(4,919)HEAD
          READ(5,600)NCCONC
          WRITE(4,600)NCCONC
          READ(5,601)(DISTC(I),CCC(I), I = 1,NCCONC)
          WRITE(4,601)(DISTC(I),CCC(I),I = 1,NCCONC)
          DO 1653 I = 1,NMESH
          XI = X(I)
          CALL INTERP(DISTC,CCC,XI,CCONC(I),NCCONC,1)
          CCONC(I) = 10.0**(CCONC(I))
1653      CONTINUE
```

```
             READ(5,901)HEAD
             WRITE(4,919)HEAD
             READ(5,901)HEAD
             WRITE(4,919)HEAD
             READ(5,600)NDEPTH
             WRITE(4,600) NDEPTH
             READ(5,601) (DDIST(I), DEPTH(I),I = 1, NDEPTH)
             WRITE(4,601) (DDIST(I), DEPTH(I),I = 1,NDEPTH)
             READ(5,901)HEAD
             WRITE(4,919)HEAD
     20      READ(5,600) NWIDTH
             WRITE(4,600)NWIDTH
             READ(5,601)(WDIST(I),WIDTH(I),I = 1,NWIDTH)
             WRITE(4,601)(WIDST(I),WIDTH(I),I = 1,NWIDTH)
             DO 680 I = 1,NMESH
             XI = X(I)
             CALL INTERP(WDIST, WIDTH, XI,W(I),NWIDTH,1)
     680     CONTINUE
             READ(5,901)HEAD
             WRITE(4,919)HEAD
             READ(5600)NFACT
             WRITE(4,600)NFACT
             READ(5,601)(DFDIS(I),DFACT(I),I = 1,NFACT)
             WRITE(4,601) (DFDIS(I),DFACT(I),I = 1,NFACT)
     600     FORMAT(I3)
     601     FORMAT(2E10.3)
             DO 1100 I = 1, NMESH
             XI = X(I)
             CALL INTERP(DDIST,DEPTH,XI,D(I),NDEPTH,1)
             CALL INTERP(DFDIS,DFACT,XI,DF(I),NFACT,1)
     1100    CONTINUE
     C
             READ(5,901)HEAD
             WRITE(4,919)HEAD
     C       SET PLANT BIOMASS
     C
             READ(5,901)HEAD
             WRITE(4,919)HEAD
             READ(5,600) NBIOMS
             WRITE(4,600) NBIOMS
             READ(5,601) (DISTBM(I),BIOMS(I),I = 1,NBIOMS)
             WRITE (4,601) (DISTBM(I),BIOMS(I),I = 1,NBIOMS)
             DO 1700 I = 1, NMESH
             XI = X(I)
             CALL INTERP(DISTBM,BIOMS,XI,BIO(I),NBIOMS,1)
```

```
1700        CONTINUE
C
C
C           SET BENTHAL RESPIRATION COEFFICIENT
C
            READ(5,901)HEAD
            WRITE(4,919)HEAD
            READ(5,600)NBENR
            WRITE(4,600)NBENR
            READ(5,601)(DISTBR(I),BENR(I),I = 1,NBENR)
            WRITE(4,601)(DISTBR(I),BENR(I),I = 1,NBENR)
            DO 1800 I = 1,NMESH
            XI = X(I)
            CALL INTERP(DISTBR,BENR,XI,BRES(I),NBENR,1)
1800        CONTINUE
C
C           CALC SATURATED OXYGEN
            NDOT = 6
            DOT(1) = 14.60
            DOT(2) = 12.80
            DOT(3) = 11.30
            DOT(4) = 10.20
            DOT(5) = 9.20
            DOT(6) = 8.40
            DTEMP(1) = 0.0
            DTEMP(2) = 5.0
            DTEMP(3) = 10.0
            DTEMP(4) = 15.0
            DTEMP(5) = 20.0
            DTEMP(6) = 25.0
            CALL INTERP(DTEMP,DOT,TEMP,CS,NDOT,1)
C
C
            READ(5,901)HEAD
            WRITE(4,919)HEAD
            READ(5,600) NFL
            WRITE(4,600) NFL
            READ(5,601)(FLDIST(I),FL(I),I = 1,NFL)
            WRITE(4,601)(FLDIST(I),FL(I),I = 1,NFL)
            DO 875 I = 1, NFL
            FL(I) = FL(I)*3600.0
875         CONTINUE
            FLT = 0.0
            J = 1
            NNN = NMESH − 1
```

```
            DO 800 I = 1, NNN
            IF(X(I).LT.FLDIST(J)) GOTO 850
            FLT = FL(J) + FLT
            J = J + 1
            Q(I) = FLT
            GOTO 800
850         Q(I) = FLT
800         CONTINUE
            Q(NMESH) = Q(NMESH − 1)
            DO 900 I = 1, NMESH
            A(I) = W(I)*D(I)
            U(I) = Q(I)/A(I)
900         CONTINUE
C
            V1C(1) = V1M
            DO 22 I = 2, NMESH
            V1C(I) = V1C(I − 1) + DX*(A(I − 1) + A(I))/2.0
22          CONTINUE
            OMEGA = 2.0*3.143/12.4
C           CALC DISPERSION COEFF AND STABILITY CRITERIA
C           DZ = DAMPL*SIN(OMEGA*( − DT))
            DO 23 I = 1, NMESH
            DOLD(I) = D(I) + DZ*DF(I)
            A(I) = DOLD(I)*W(I)
23          CONTINUE
            VC(1) = V1M + DZ*DF(1)*W(1)*ULEN/2.0
            DO 24 I = 2, NMESH
            VC(I) = VC(I − 1) + DX*(A(I − 1) + A(I))/2.0
            I(I) = (Q(I) + (VC(I) − V1C(I))/DT)/A(I)
24          CONTINUE
C
            DO 1200 I = 1,NMESH
            UU(I) = ABS(U(I))
            E(I) = 22.6*UU(I)*AMAN*((D(I)/0.304)**0.833)
            E(I) = E(I) + DISP
1200        CONTINUE
1190        DO 1007 I = 1,NMESH
            SIG(I) = E(I)*DT/(DX*DX)
            CR(I) = U(I)*DT/DX
            AA = 3.0 − 2.0*CR(I)
            BB = 1.0 − CR(I)**2
            DD = 6.0*(1.0 − 2.0*CR(I))
            EE = CR(I)**2 − 1.0
            FF = 6.0*(2.0*CR(I) − 1.0)
            Z1 = AA*BB/DD
```

```
            Z2 = AA*EE/FF
            IF(CR(I).GE.1.5) GOTO 999
            IF (CR(I).LT.0.5.AND.SIG(I).LE.Z1) GOTO 1007
            IF (CR(I).GT.0.5.AND.CR(I).LT.1.0) GOTO 999
            IF (CR(I).GT.0.5.AND.SIG(I).GE.Z2) GOTO 1007
 999        WRITE(4,605)
 605        FORMAT(27HSTABILITY CRITERIA VIOLATED)
            GOTO 2050
1007        CONTINUE
C
C
C
C           READ BOD AND DO LOADS
            READ(5,901)HEAD
            WRITE(4,919)HEAD
C
            READ(5,600)NBODL
            WRITE(4,600)NBODL
            DO 1260 I = 1,NBODL
            READ(5,901)HEAD
            WRITE(4,919)HEAD
            READ(5,702)DISTBL(I)
            WRITE(4,702)DISTBL(I)
            READ(5,600)NBLD(I)
            WRITE(4,600)NBLD)(I)
            READ(5,601)(TTBOD(I,K),BL(I,K),K = 1,NBLD(I))
            WRITE(4,601)(TTBOD(I,K),BL(I,K),K = 1,NBLD(I))
1260        CONTINUE
 602        FORMAT (4E10.3)
 702        FORMAT (E10.3)
            DO 1250 I = 1,NMESH
            DO 1250 K = 1,50
            BODL(I,K) = 0.0
            BODT(I,K) = 0.0
1250        CONTINUE
            J = 1
            DO 1300 I = 1,NMESH
            IF(INT(X(I)).GE.INT(DISTBL(J))) GOTO 1310
            GOTO 1300
1310        DO 1320 K = 1,NBLD(J)
            BODL(I,K) = BL(J,K)
            BODL(I,K) = BODL(I,K)/(A(I)*DX)
            BODL(I,K) = BODL(I,K)*1000.0
            BODT(I,K) = TTBOD(J,K)
1320        CONTINUE
```

```
                 IBPOS(J) = I
                 J = J + 1
1300             CONTINUE
C
                 READ (5,901)HEAD
                 WRITE(4,919)HEAD
                 READ(5,600) NDOL
                 WRITE(4,600) NDOL
                 DO 1261 I = 1, NDOL
                 READ(5,901)HEAD
                 WRITE(4,919)HEAD
                 READ(5,702)DISTDL(I)
                 WRITE(4,702)DISTDL(I)
                 READ(5,600)NDL(I)
                 WRITE(4,600)NDL(I)
                 DO 1261 K = 1, NDL(I)
                 READ(5,601)TTDO(I,K),DL(I,K)
                 WRITE(4,601)TTDO(I,K),DL(I,K)
1261             CONTINUE
                 DO 1251 I = 1, NMESH
                 DO 1251 K = 1, 50
                 DOL(I,K) = 0.0
                 DOTT(I,K) = 0.0
1251             CONTINUE
                 J = 1
                 DO 1301 I = 1, NMESH
                 IF(INT(X(I)).GE.INT(DISTDL(J))) GOTO 1311
                 GOTO 1301
1311             DO1321 K = 1, NDL(J)
                 DOL(I,K) = DL(J,K)
                 DOL(I,K) = DOL(I,K)/(A(I)*DX)
                 DOL(I,K) = DOL(I,K)*1000.0
                 DOTT(I,K) = TTDO(J,K)
1321             CONTINUE
                 IDPOS(J) = I
                 K = J + 1
1301             CONTINUE
C                READ AMMONIA LOADS
                 READ(5,901)HEAD
                 WRITE(4,919)HEAD
                 READ(5,600)NAL
                 WRITE(4,600)NAL
                 DO 1430 I = 1, NAL
                 READ(5,901)HEAD
                 WRITE(4,919)HEAD
```

```
                READ(5,702)DISTAL(I)
                WRITE(4,702)DISTAL(I)
                READ(5,600)NALD(I)
                WRITE(4,600)NALD(I)
                READ(5,601)(TTAM(I,K),AL(I,K),K = 1,NALD(I))
                WRITE(4,601)(TTAM(I,K),AL(I,K),K = 1,NALD(I))
1430            CONTINUE
                DO 1485 I = 1,NMESH
                DO 1485 K = 1,50
                AML(I,K) = 0.0
                AMT(I,K) = 0.0
1485            CONTINUE
                J = 1
                DO 1480 I = 1,NMESH
                IF(INT(X(I)).GE.INT(DISTAL(J))) GOTO 1490
                GOTO 1480
1490            DO 1440 K = 1,NALD(J)
                AML(I,K) = AL(J,K)
                AML(I,K) = AML(I,K)/(A(I)*DX)
                AML(I,K) = AML(I,K)*1000.0
                AMT(I,K) = TTAM(J,K)
1440            CONTINUE
                IAPOS(J) = I
                J = J + 1
1480            CONTINUE
C               READ COLIFORM LOADS
                READ(5,901)HEAD
                WRITE(4,919)HEAD
                READ(5,600) NCL
                WRITE(4,600) NCL
                DO 1433 I = 1,NCL
                READ(5,901)HEAD
                WRITE(4,919)HEAD
                READ(5,702)DISTCL(I)
                WRITE(4,702)DISTCL(I)
                READ(5,600)NCLD(I)
                WRITE(4,600)NCLD(I)
                READ(5,601)(TTCL(I,K),CL(I,K),K = 1,NCLD(I))
                WRITE(4,601)(TTCL(I,K),CL(I,K)K = 1,NCLD(I))
1433            CONTINUE
                DO 1488 I = 1,NMESH
                DO 1488 K = 1,50
                CLL(I,K) = 0.0
                CLT(I,K) = 0.0
1488            CONTINUE
```

```
          J = 1
          DO 1483 I = 1,NMESH
          IF(INT(X(I)).GE.INT(DISTCL(J))) GOTO 1493
          GOTO 1483
1493      DO 1443 K = 1,NCLD(J)
          CLL(I,K) = 10.0**(CL(J,K))
          CLL(I,K) = CLL(I,K)/10000.0
          CLL(I,K) = CLL(I,K)/(A(I)*DX)
          CLT(I,K) = TTCL(J,K)
1443      CONTINUE
          ICPOS(J) = I
          J = J + 1
1483      CONTINUE
C         SET PLANT BIOMASS
C
C         SET BOD COEFFICIENT
C
C
C         ADJUST COEFFS FOR TEMP
C
          PW = TEMP − 20.0
          AK1 = AK1*1.047**PW
          AK5 = AK5*1.047**PW
          AK6 = AK6*1.08**PW
          AK9 = AK9*1.05**PW
          PMAX = PMAX*1.02**PW
          DO 50 I = 1,NMESH
          BRES(I) = BRES(I)*1.047**PW
50        CONTINUE
          DO 40 I = 1,NMESH
          WRITE(4,651) X(I),CDO(I),CBOD(I),ACONC(I),S(I),
            CCONC(I)
631       FORMAT(I3)
40        CONTINUE
C
C         CALCULATION OF DO
C
          T = TSTART
          TCOUNT = 0.0
2000      T = T + DT
C         DZ = DAMPL*SIN(OMEGA*T)
          THT = T − THW
          IF(THT.LT.0.0) THT = THT + 12.4
          IF(THT.GT.24.8) THT = MOD(THT,12.4)
          CALL INTERP(TX,TY,THT,DZ,MTIDE,NINT)
```

```
            DO 1810 I = 1,NMESH
            DNEW(I) = D(I) + DZ*DF(I)
            A(I) = DNEW(I)*W(I)
1810        CONTINUE
            V1C(1) = V1M + DZ*DF(1)*W(1)*ULEN/2.0
            DO 1805 I = 2,NMESH
            V1C(I) = V1C(I − 1) + DX*(A(I − 1) + A(I))/2.0
            U(I) = (Q(I) + (VC(I) − V1C(I))/DT)/A(I)
            UU(I) + ABS(U(I))
1805        CONTINUE
            CBOD(1) = UBOD + AMPB*(SIN(6.284*T/PERB))
            CDO(1) = UDO + AMPD*(SIN(6.284*T/PERD))
            ACONC(1) = UAM + AMPA*(SIN(6.284*T/PERA))
            S(1) = UAS + AMPS*(SIN(6.284*T/PERS))
            CCONC(1) = UC + AMPC*(SIN(6.284*T/PERC))
            CBOD(NMESH) = DBOD +
               DAMPB*(SIN(6.284*T/DPERB))
            CDO(NMESH) = DDO + DAMPD*(SIN(6.284*T/DPERD))
            ACONC(NMESH) = DAM + DAMPA*(SIN(6.284*T/
               DPERA))
            S(NMESH) = DS + DAMPS*(SIN(6.284*T/DPERS))
            CCONC(NMESH) = DC + DAMPC*(SIN(6.284*T/DPERC))
            TCOUNT = TCOUNT + DT
            DO 1820 I = 1,NMESH
            CALL SORCD(T,PD(I),I,NDOL)
            CALL SINKD(RD(I),I)
            C(I) = CDO(I)
1820        CONTINUE
            NS = 1
            IF(TCOUNT.EQ.DT) NS = 2
            CALL QUICK(E,PD,RD,C,F,NMESH,NS)
            DO 1850 I = 1,NMESH
            CDO(I) = F(I)
1850        CONTINUE
            DO 1870 I = 1,NMESH
            CALL SORCB(T,PB(I),I,NBODL)
            CALL SINKB(RB(I),I)
            C(I) = CBOD(I)
1870        CONTINUE
            CALL QUICK(E,PB,RB,C,F,NMESH,NS)
            DO 1895 I = 1,NMESH
            CBOD(I) = F(I)
1895        CONTINUE
C
C           CALC AMMONIA
```

```
C
              DO 1880 I = 1,NMESH
              CALL SORCA(T,PA(I),I,NAL)
              CALL SINKA(RA(I),I)
              C(I) = ACONC(I)
1880          CONTINUE
              CALL QUICK(E,PA,RA,C,F,NMESH,NS)
              DO 1890 I = 1,NMESH
              ACONC(I) = F(I)
1890          CONTINUE
C
C             CALC SALINITY
              DO 1897 I = 1,NMESH
              C(I) = S(I)
1897          CONTINUE
              CALL QUICK(E,DUMY,DUMY,C,F,NMESH,NS)
              SMASS = 0.0
              DO 1896 I = 1,NMESH
              VC(I) = V1C(I)
              S(I) = F(I)
              SMASS = SMASS + S(I)*A(I)*DX
1896          CONTINUE
C
C             CALC COLIFORMS
C
              DO 1883 I = 1,NMESH
              CALL SORCC(T,PC(I),I,NCL)
              CALL SINKC(T,RC(I),I)
              C(I) = CCONC(I)
1883          CONTINUE
              DO 432 I = 1,100
432           CONTINUE
              CALL QUICK(E,PC,RC,C,F,NMESH,NS)
              DO 1893 I = 1,NMESH
              CCONC(I) = F(I)
1893          CONTINUE
C
              IF(INT(TCOUNT*10.0).EQ.INT(TPRNT*10.0)) GOTO 1840
              GOTO 2000
1840          TIME = T
              WRITE(4,650) T
              WRITE(6,650) T
650           FORMAT(1PE10.3)
C             WRITE(4,652)SMASS
C             WRITE(6,652)T,SMASS
```

```
652          FORMAT(3E10.3)
             DO 1860 I = 1,NMESH
             WRITE(4,651) X(I),CDO(I),CBOD(I),U(I),S(I),.CCONC(I)
             WRITE(6,651) X(I),CDO(I),CBOD(I),U(I),S(I),.CCONC(I)
651          FORMAT(1P6E11.3)
1860         CONTINUE
             WRITE(4,652)T,U(40),U(55)
             WRITE(4,650) T
C            WRITE(6,651) T,S(10),S(30),S(60),S(80)
C            WRITE(6,651) U(10),U(30),U(60),U(80)
             TCOUNT = 0.0
             IF(T.LT.TSTOP) GOTO 2000
2050         STOP
             END
C
C
             SUBROUTINE SORCD (T,P,I,NDOL)
             COMMON /DOX/CS,AK2,PDO,(101),ADO(101)
             COMMON /DOX1/D(101),CDO(101),DOL(101,50),
                    DOTT(101,50)
             COMMON /VEL/U(101),Q(101),A(101),W(101)
             COMMON /ILPOS/IBPOS(20),IAPOS(20),IDPOS(20),
                    ICPOS(20)
             COMMON /NLOD/NBLD(20),NDL(20),NALD(20),NCLD(20)
             DIMENSION Z(101),Y(101)
C
C
             CALL REAER(P1,I)
             CALL PSYN(T,P2,I)
C
             DO 700 J = 1,NDOL
             M = J
             IF(IDPOS(J).EQ.I) GOTO 701
             DL = 0.0
700          CONTINUE
             GOTO 702
C
701          DO 20 J = 1,NDL(M)
             Z(J) = DOTT(I,J)
             Y(J) = DOL(I,J)
20           CONTINUE
             CALL INTERP(Z,Y,T,DL,NDL(M),1)
702          P = P1 + P2 + DL
             RETURN
             END
```

```
C
            SUBROUTINE REAER(P1,I)
            COMMON /DOX/CS,AK2,PDO(101),ADO(101)
            COMMON /DOX1/D(101),CDO(101),DOL(101,50),
                    DOTT(101,50)
            COMMON /VEL/U(101),Q(101),A(101),W(101)
            COMMON /U/UU(101)
            AK2 = ((21.7*(UU(I)/1094.4)**0.67)/((D(I)/0.304)**1.85))/24.0
            P1 + AK2*(CSA − CDO(I))*A(I)
            RETURN
            END
C
            SUBROUTINE PSYN(T,P2,I)
            COMMON /PHOT/BIO(101),PMAX,DAYLEN,TS
            COMMON /DOX1/D(101),CDO(101),DOL(101,50),
               DOTT(101,50)
            COMMON /VEL/U(101),Q(101),A(101),W(101)
            IF(T.LT.TS + DAYLEN) GOTO 10
            TS = TS + 24.0
     10     IF(T.LE.TS) GOTO 20
            IF(T.GE.TS + DAYLEN) GOTO 20
            P2 = PMAX*SIN((3.142/DAYLEN)*(T − TS))
            P2 = (P2*BIO(I))*A(I)/D(I)
            RETURN
     20     P2 = 0.0
            RETURN
            END
C
C
            SUBROUTINE SINKD (R,I)
C
C
            CALL CONBOD(R1,I)
            CALL BENRSP(R2,I)
            CALL PLRSP(R3,I)
            CALL NITRIF(R4,I)
            R = R1 + R2 + R3 + R4*4.33
            RETURN
            END
C
            SUBROUTINE CONBOD(R1,I)
            COMMON /DOX1/D(101),CDO(101),DOL(101,50),
               DOTT(101,50)
            COMMON /VEL/U(101),Q(101),A(101),W(101)
            COMMON /COEF/AK1,AK4,AK5,AK6
```

```
                COMMON /BOX1/CBOD(101),BODL(101,50)BODT(101,50)
                COMMON /U/UU(101)
                AK4 = 1.0
                AKD = AK1 + AK4*UU(I)/(D(I)*3600.0*24.0)
                R1 = AKD*CBOD(I)*A(I)
                RETURN
                END
C
                SUBROUTINE BENRSP(R2,I)
                COMMON /DOX1/D(101),CDO(101),DOL(101,50),
                        DOTT(101,50)
                COMMON /VEL/U(101),Q(101),A(101),W(101)
                COMMON /RESP/BRES(101)
                R2 = BRES(I)*A(I)/D(I)
                RETURN
                END
C
                SUBROUTINE PLRSP(R3,I)
                COMMON /PHOT/BIO(101),PMAX,DAYLEN,TS
                COMMON /DOX1/D(101),CDO(101),DOL(101,50),
                    DOTT(101,50)
                COMMON /VEL/U(101),Q(101)A(101),W(101)
                COMMON /COEF/AK1,AK4,AK5,AK6
                R3 = BIO(I)*AK5*A(I)/D(I)
                RETURN
                END
C
C
                SUBROUTINE NITRIF(R4,I)
                COMMON /VEL/U(101),Q(101),A(101),W(101)
                COMMON /COEF/AK1,AK4,AK5,AK6
                COMMON /AMON/ACONC(101),AML(101,50),AMT(101,50)
                R4 = AK6*ACONC(I)*A(I)
                RETURN
                END
C
C
C
                SUBROUTINE QUICK(E,P,R,C,F,N,NS)
                COMMON /VEL/U(101),Q(101),A(101),W(101)
                COMMON /DOX1/D(101),CDO(101),DOL(101,50),
                    DOTT(101,50)
                COMMON /MESH/DX,DT
                DIMENSION E(N),P(N),R(N),C(N)
                DIMENSION AC(101),F(101),DP(101),AN1(101)
```

```
          DIMENSION GF(101),UR(101),CR(101),AR(101)
          DIMENSION CVR(101),CV(101),PI1(101),PI2(101)
C
          GOTO (8,10),NS
C
   10     Z = DT/DX
          NM1 = N − 1
          NM2 = N − 2
C
    8     DO 11 I = 1,N
          AC(I) = A(I)*C(I)
          DP(I) = E(I)*DT/DX/DX
          P(I) = P(I)/A(I)
          R(I) = R(I)/A(I)
   11     CONTINUE
C
          DO 12 I = 1,NM2
   12     GF(I) = (C(I + 2) − C(I + 1))/DX
          GF(NM1) = GF(NM2)
          GF(N) = GF(NM1)
          DO 13 I = 2,N
   13     CVR(I) = (GF(I) − GF(I − 1))/DX
          CVR(1) = CVR(2)
C
          DO 14 I = 2,N
          IF(I.EQ.N) GOTO 21
          UR(I) = (U(I + 1) + U(I))/2.0
          AR(I) = (A(I + 1) + A(I))/2.0
          GOTO 22
   21     UR(I) = UR(I − 1)
          AR(I) = AR(I − 1)
   22     CONTINUE
          CR(I) = UR(I)*Z
          IF(UR(I).GE.0.0) CV(I) = CVR(I − 1)
          IF(UR(I).LT.0.0) CV(I) = CVR(I)
          IF(I.EQ.N) GOTO 23
          BA = 0.5*(C(I) + C(I + 1))
          GOTO 24
   23     BA = 0.5*(C(I = 1) + C(I))
   24     CONTINUE
          BB = − DX/2.0*CR(I)*GF(I − 1)
          BC = DX*DX/2.0*(DP(I) − 1.0/3.0*(1.0 − CR(I)*CR(I)))*CV(I)
          PI1(I) = BA + BB + BC
          PI2(I) = GF(I − 1) − DX/2.0*CR(I)*CV(I)
   14     CONTINUE
```

```
                UR(1) = (U(2) + U(1))/2.0
                PI1(1) = (C(2) + C(1))/2.0 − 0.5*UR(1)*Z*(C(2) − C(1))
                PI2(1) = PI2(2)
C
                DO 2 I = 2,N
                AN1(I) = A(I) + Z*(UR(I − 1)*AR(I − 1) − UR(I)*AR(I))
     2          CONTINUE
                AN1(1) = AN1(2)
C
                DO 1 I = 2,N
                F(I) = AC(I)/AN1(I)
                F(I) = F(I) + Z*(UR(I − 1)*PI1(I − 1)*AR(I − 1) −
                   UR(I)*PI1(I) *AR(I))/AN1(I)
                F(I) = F(I) + Z*( − E(I − 1)*PI2(I − 1) +
                   E(I)*PI2(I))*A(I)/AN1(I)
                F(I) = F(I) + DT*(P(I) − R(I))*A(I)/AN1(I)
                IF(F(I).LT.0.0) F(I) = 0.0
     1          CONTINUE
                F(1) = C(1)
                F(N) = C(N)
C
                RETURN
                END
C
C
                SUBROUTINE SORCB(T,P,I,NBODL)
C
                COMMON /VEL/U(101),Q(101),A(101),W(101)
                COMMON /BOX1/CBOD(101),BODL(101,50),BODT(101,50)
                COMMON /ILPOS/IBPOS(20),IAPOS(20),IDPOS(20),
                   ICPOS(20)
                COMMON /NLOD/NBLD(20),NDL(20),NALD(20),NCLD(20)
                DIMENSION Z(101),Y(101)
C
                DO 700 J = 1,NBODL
                M = J
                IF(IBPOS(J).EQ.I) GOTO 701
                BL = 0.0
    700         CONTINUE
                GOTO 702
C
    701         DO 20 J = 1,NBLD(M)
                Z(J) = BODT(I,J)
                Y(J) = BODL(I,J)
     20         CONTINUE
```

```
         CALL INTERP(Z,Y,T,BL,NBLD(M),1)
702      P = BL
         RETURN
         END
C
C
C
C
C
         SUBROUTINE SINKB(R,I)
         CALL CONBOD(R1,I)
         R = R1
         RETURN
         END
C
C
         SUBROUTINE SORCA(T,P,I,NAL)
         COMMON /VEL/U(101),Q(101),A(101),W(101)
         COMMON /AMON/ACONC(101),AML(101,50),AMT(101,50)
         COMMON /NLOD/NBLD(20),NDL(20),NALD(20),NCLD(20)
C
         COMMON /ILPOS/IBPOS(20),IAPOS(20),IDPOS(20),
           ICPOS(20)
         DIMENSION Z(101),Y(101)
C
         DO 700 J = 1,NAL
         M = J
         IF(IAPOS(J).EQ.I) GOTO 701
         AL = 0.0
700      CONTINUE
         GOTO 702
C
701      DO 20 J = 1,NALD(M)
         Z(J) = AMT(I,J)
         Y(J) = AML(I,J)
20       CONTINUE
         CALL INTERP(Z,Y,T,AL,NALD(M),1)
702      P = AL
         RETURN
         END
C
C
         SUBROUTINE SINKA(R,I)
         CALL NITRIF(R4,I)
         R = R4
```

```
            RETURN
            END
C

            SUBROUTINE SORCC(T,P,I,NCL)
            COMMON /VEL/U(101),Q(101),A(101),W(101)
            COMMON /COLI/CCONC(101),CLL(101,50),CLT(101,50)
            COMMON /NLOD/NBLD(20),NDL(20),NALD(20),NCLD(20)
C

            COMMON /ILPOS/IBPOS(20),IAPOS(20),IDPOS(20),
              ICPOS(20)
            DIMENSION Z(101),Y(101)
C

            DO 700 J = 1,NCL
            M = J
            IF(ICPOS(J).EQ.I) GOTO 701
            CL = 0.0
  700       CONTINUE
            GOTO 702
C
  701       DO 20 J = 1,NCLD(M)
            Z(J) = CLT(I,J)
            Y(J) = CLL(I,J)
   20       CONTINUE
            CALL INTERP(Z,Y,T,CL,NCLD(M),1)
  702       P = CL
            RETURN
            END
C
C

            SUBROUTINE SINKC(T,R,I)
            COMMON /COLI/CCONC(101),CLL(101,50),CLT(101,50)
            COMMON /CHL/TTCL(20,50),CL(20,50)
            CALL DIEOFF(T,R4,I)
            R = R4
            RETURN
            END
C

            SURBOUTINE DIEOFF(T,R4,I)
            COMMON /COLI/CCONC(101),CLL(101,50),CLT(101,50)
            COMMON /CHL/TTCL(20,50),CL(20,50)
            COMMON /PHOT/BIO(101),PMAX,DAYLEN,TS
            COMMON /VEL/U(101),Q(101),A(101),W(101)
            IF(T.LT.TS + DAYLEN) GOTO 10
            TS = TS + 24.0
   10       IF(T.LE.TS) GOTO 20
```

```
            IF(T.GE.TS + DAYLEN) GOTO 20
            AK9 = AK9*SIN((3.142/DAYLEN)*(T − TS))
            R4 = AK9*CCONC(I)
            RETURN
      20    R4 = 0.0
            RETURN
            END
C

            SUBROUTINE INTERP(X,Y,X0,Y0,M,NINT)
            DIMENSION X(M),Y(M)
            GOTO (1,2),NINT
C
      1     K1 = 1
            K2 = M
      4     K = (K1 + K2)/2
            IF(X0 − X(K))6,7,8
      6     K2 = K
            IF(K2 − K1 − 1)9,9,4
      7     K1 = K
            GOTO 9
      8     K1 = K
            IF(K2 − K1 − 1)9,9,4
      9     PROP = (X0 − X(K1))/(X(K1 + 1) − X(K1))
      2     Y0 = Y (K1) + PROP*(Y(K1 + 1) − Y(K1))
            RETURN
            END
C
```

GLOSSARY OF NAMES

TLEN	= Total length being modelled	(m)
DX	= Distance step	(m)
DT	= Time step	(h)
UMAX	= Maximum velocity	(m/h)
NMESH	= Number of mesh points along length	
X(I)	= Mesh distance	(m)
NDEPTH	= Number of depth readings	
DDIST	= Distance downstream of depth reading	(m)
D(I)	= Interpolated depth	(m)

A(I)	= Interpolated cross-sectional area at distance I	(m^2)
INTERP	= Interpolation subroutine	
NVEL	= Number of readings of velocity	
VDIST	= Distance from upstream boundary of velocity measurement	(m)
VEL	= Velocity reading	(m/h)
NFL	= Number of points at which there are flow readings	
FLDIST	= Distance from upstream boundary of flow measurement	(m)
FL	= Flow reading	(m^3/h)
Q(I)	= Interpolated flow at each mesh point	(m^3/h)
U(I)	= Interpolated velocity	
AMAN	= Manning's roughness coefficient	
DISP	= Dispersion constant	(m^2/h)
E	= Dispersion coefficient	(m^2/h)
SIG	$= \dfrac{E\partial^2 t}{\partial x^2}$	
CR	= Courant number $\dfrac{U\partial t}{\partial x}$	
AA, BB, DD, EE, FF, Z1, Z2	are intermediate parameters in the calculation of the stability of the numerical scheme.	
BODL	= BOD load	(g/h)
DOL	= DO load	(g/h)
BNODL	= Number of BOD loads	
DISTBL	= Distance of BOD discharge from upstream boundary	(m)
BL	= BOD load	(g/h)
NDOL	= Number of DO loads	
DISTD	= Distance of DO discharge from upstream boundary	(m)
DL	= DO load	(g/h)
NBOD	= Number of readings of BOD concentration	
DISTB	= Distance of BOD reading from upstream boundary	(m)
BOD	= BOD reading	(g/m^3)
CBOD	= Interpolated BOD	(g/m^3)
NDO	= Number of DO readings	
DISTDO	= Distance of DO readings from upstream boundary	(m)

DO	= DO reading	(g/m^3)
CDO	= Interpolated DO	(g/m^3)
UBOD	= BOD at upstream boundary	(g/m^3)
UDO	= DO at upstream boundary	(g/m^3)
UAM	= Ammonia concentration at upstream boundary	(g/m^3)
NBIOMS	= Number of readings of plant biomass	
DISTBM	= Distance of biomass reading from upstream boundary	(m)
BIOMS	= Dry weight of plant biomass readings	(kg dry weight/m^2)
BIO	= Interpolated biomass	(kg dry weight/m^2)
NAL	= Number of ammonia loads	
DISTAL	= Distance of ammonia loads	(m)
Al	= Ammonia load	(g/h)
CS	= Saturation concentration for dissolved oxygen	(g/m^3)
AK_4	= Coefficient of resuspension	
NBENR	= Number of benthal respiration readings	
DISTBR	= Distance of benthal respiration reading from upstream boundary	(m)
BENR	= Benthal respiration reading	(g O_2/m^2 h)
BRES	= Interpolated benthal respiration	
AKI	= BOD coefficient	(base e per hour)
AK5	= Plant respiration coefficent	(g O_2/kg dry weight/h)
PMAX	= Maximum rate of photosynthesis	(g O_2/kg dry weight/h)
DAYLEN	= Length of daylight	(h)
TS	= Time of sunrise	(h)
P	= Source term (for DO also called PD, for BOD called PB and for ammonia, PA)	
R	= Sink term (also RD, RB & RA on the same basis)	
TSTART	= Time of start of computer run	(h)
TSTOP	= Time of stop of computer run	(h)
TPRNT	= Print time	(h)
TCOUNT	= Counter for TPRNT	(h)
QUICK	= Finite difference scheme for calculating rates of change of DO and BOD	
SOURCD SINKD }	= Sources and sinks subroutines for DO	

SOURCB SINKB	= Sources and sinks subroutines for BOD	
RAEAR	= Subroutines for re-aeration	
PSYNT	= Subroutine for photosynthesis	
CONBOD	= Subroutine for BOD decay	
BENRESP	= Subroutine for benthal respiration	
PLRSP	= Subroutine for plant respiration	
NWEIR	= Number of weirs	
AWEIR BWEIR	= Coefficients (see notes)	
HWEIR	= Height of weir	(m)
NDOT	= Number of DO saturation values	
DOT	= DO saturation at a specific temperature	(g/m^3)
DTEMP	= Temperature of DO saturation	$(°C)$

Chapter 9

Modelling Water Quality in Lakes and Reservoirs

A. JAMES

9.1 INTRODUCTION

Modelling the quality of water in lakes and reservoirs is very different to modelling quality changes in rivers and estuaries or even the sea. The main uses of water in lakes and reservoirs are similar to those in rivers, namely amenity, fisheries and abstraction for supply, but the lotic environment of a lake or reservoir creates important differences, from those of the lentic environment of a stream. The main differences are as follows:

(a) Lakes and reservoirs rarely receive discharges of organic matter large enough to cause serious oxygen depletion.

(b) Lakes and reservoirs have much greater retention times than most rivers. This enables planktonic algae to become dominant and makes them sensitive to enrichment by inorganic nutrients.

(c) The response time of lakes and reservoirs to pollution is much greater than that of streams, both in showing the effects of pollution and in recovery.

(d) The principal water quality gradients are in the vertical direction rather than the longitudinal direction.

As a consequence of these differences, the majority of lake models are concerned with thermal stratification and eutrophication. Quality considerations relating to BOD/DO are of secondary importance. Similarly the main kinetics of interest are algal rather than bacterial. The differences in modelling approach also reflect the main

An Introduction to Water Quality Modelling. 2nd Edition. Edited by A. James
© 1993 John Wiley & Sons Ltd.

factors which determine the quality of water in lakes and reservoirs which are due mainly to the following factors in various combinations:

(a) influent quality;
(b) mixing pattern;
(c) physical and chemical processes during storage;
(d) biological growths and their role in the removal and release of substances.

Modelling of water quality changes in lakes therefore involves representation of all of the above, although not necessarily in a single model. Depending upon the aims and the processes represented, lake models can be conveniently classified as shown in Table 9.1. The classes shown there are discussed below.

9.2 PHYSICAL MODELS

As indicated in Table 9.1 the main interest in the physical aspects of limnology are related to temperature models, either of the overall temperature in unstratified lakes or of the thermal structure in stratified lakes.

Other models of the fluid dynamics of lakes have been used to provide the basic hydrodynamics for water quality models. This approach has been less often used since many smaller lakes are well mixed laterally and longitudinally and even large lakes may often be represented as a set of embayments.

All three types of physical model are described below.

9.2.1 Modelling of Overall Temperature

As shown in Figure 9.1 it is possible to calculate the temporal temperature pattern in a lake using a heat budget approach. Such models are useful where water bodies (natural or artificial) are used for cooling water. The calculation of the sources and sinks are discussed in detail in Chapter 7 for temperature models of rivers. In the case of lake models the sources and sinks are often simplified, e.g.

$$E + H = (a + b)(CVP_0 - P_a) + 0.6(T_0 - T_a) \tag{9.1}$$

Table 9.1 Classification of the main types of lake model

Type of model		Main aim	
		Planning	Operational
Physical	Temperature	$\checkmark$	$\checkmark$
	Thermal structure	$\checkmark$	$\checkmark$
	Hydrodynamic	$\checkmark$	$\checkmark$
Chemical	Mass balance loading	$\checkmark$	
Biological	Eutrophication	$\checkmark$	$\checkmark$
	Toxicity	$\checkmark$	
	(food chain)		

($\checkmark$ indicates commonly applied type of model)

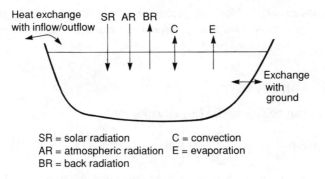

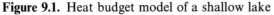

SR = solar radiation   C = convection
AR = atmospheric radiation E = evaporation
BR = back radiation

Figure 9.1. Heat budget model of a shallow lake

where E = evaporative loss
      H = exchange with atmosphere
      VP_0 & VP_a = vapour pressure (saturation and actual)
      T_0 & T_a = temperature (water and air)
      a & b are coefficients

9.2.2 Models of Thermal Structure

Stratification is of crucial importance in determining water quality in lakes and it is therefore important to be able to predict temperature patterns. To some extent this can be determined from the latitude and altitude, as shown in Figure 9.2.

But depth and wind also play an important part, especially in lakes of intermediate depth between 10 m and 30 m. Shallow lakes (10 m) rarely stratify for long periods and deep lakes (30 m) generally stratify on a longer-term basis, either permanently in tropical climates or seasonally outside the tropics.

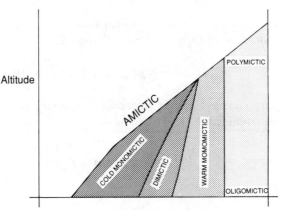

Figure 9.2. Thermal patterns in lakes

Where stratification is likely to occur, it is useful to be able to predict the temperature distribution. This can be done using either of the following models.

Eddy Diffusivity Model

This approach can be summarized by equation (9.2):

$$A(z)\left[\frac{\partial T}{\partial t} + W\frac{\partial T}{\partial z}\right] = \frac{\partial}{\partial z}\left[A(z)(\alpha - K_H)\frac{\partial T}{\partial z}\right] + \frac{\partial(A\phi(z))/\partial z}{\rho C_p} \qquad (9.2)$$

| Rate of change of temperature (vertically) | Advective transfer of heat | Molecular & eddy diffusion transfer of heat | Heat source |

where A = cross-sectional area
     T = temperature
     t = time
     z = depth
     W = velocity in the z direction
     α = coefficient of molecular diffusion
     K_H = coefficient of eddy diffusion
     ϕ = flux of solar radiation
     ρ = density of water
     C_p = specific heat of water

When $A(z)$ is constant and vertical advection is neglected, equation (9.2) can be simplified to give

$$\frac{\partial T}{\partial t} = \frac{\partial}{\partial z}\left[(\alpha + K_H)\frac{\partial T}{\partial z}\right] - \frac{\partial\phi/\partial z}{\rho C_p} \qquad (9.3)$$

The heat transfer equation may be integrated from depth z to the bottom z_b, to express the results in terms of the temperature change, θ:

$$\frac{\partial}{\partial t}\int_z^{z_b} \rho C_p A\theta \, du = -\rho C_p A(\alpha + K_H)\frac{\partial\theta}{\partial z} + AR \qquad (9.4)$$

When ρ and C_p are constant and R (incoming radiation) is integrated over all wavelengths, the equation can be re-written as

$$(\alpha + K_H) = -\frac{1}{\partial\theta/\partial z}\left(\frac{S}{A} - \frac{R}{C_p}\right) \qquad (9.5)$$

where S is given by

$$S = \frac{d}{dt}\int_z^{z_b} A\theta \, d\mu$$

In the absence of data on the value of K_H it is important to appreciate that its value depends upon the water velocity, which in turn depends upon wind speed and friction.

The effect of molecular diffusion may be neglected and the diffusive term can be expressed as

$$K_H = \frac{1}{\rho}\left(\frac{W^2}{\partial v/\partial z}\right)(1 + \alpha S)^{-h} \tag{9.6}$$

Where the friction velocity, v, is related to the wind speed, W. The stability parameter, S, may be expressed in the form of a Richardson number:

$$S = [-gK^2(Z_0 - z)^2 \rho W^2]\frac{\partial e}{\partial z} \tag{9.7}$$

where K is the Von Karman coefficient
     P is the Prandtl number
     α and n are coefficients

When the radiation term $\phi(z)$ is included in the eddy diffusion model, then the following two criteria must be satisfied by the time step Δt to ensure stability:

$$\Delta t \le \frac{(\Delta z)^2}{z(\alpha + K_H)}$$

$$\Delta t \le \frac{\Delta z}{W}$$

where W is the vertical velocity.

Energy Budget Model

In this approach, the vertical temperature profile is assumed to be composed of two parts as shown in Figure 9.3.

The completely mixed upper part is assumed to consist of m layers of unit thickness, all of which are isothermal. Below is a series of n layers of the same thickness, in which the temperature and hence the density are changing. Any turbulent energy generated by wind or convection is dissipated in increasing the depth of the mixed layer. At each time step, a comparison is made between the mechanical energy available and the potential energy required to deepen the mixed layer. E_k is the wind energy over the last time step

$$E_k{:}E_p \tag{9.8}$$

$$E_k = \tau W^* A_s \Delta t$$

where A_s = surface area of the lake
    τ = friction factor
    Δ = shear velocity

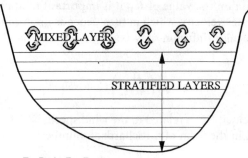

For Ratio $E_k : E_p < 1$

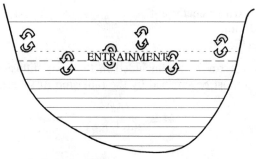

For Ratio $E_k : E_p \geq 1$

$$E_p = \sum_{i=1}^{n} V(i, k) [\rho(m + 1, k) - \rho(i, k)] [m + 1 - i] \Delta x$$

$$E_k = \tau \, w^* \, A \, \Delta t$$

$$V(i, k) = Volume$$

$$\rho(i, k) = Density$$

Figure 9.3. Mixed layer or energy budget model

E_p is the potential energy required to lift the layer of colder water into the middle of the mixed zone.

$$E_p = g \sum_{i=1}^{n} V(i, k)[p(m + 1, k) - P(i, k)][m + 1 - i] \Delta z \qquad (9.9)$$

where $V(i, k)$ = volume of layer i at time of k
$P(i, k)$ = density of layer i at time of k

At each time step the input of wind energy is used to increase the potential energy of the mixed layer by entrainment provided that $E_k > E_p$. When $E_k < E_p$ the wind energy is assumed to be dissipated in overcoming viscosity and no increase in the thickness of the mixed layer takes place.

The total mixing energy E_t may be calculated from

$$E_t = E_k + E_c \tag{9.10}$$

where

$$E_c = \begin{cases} 0.1 \dfrac{Q_n A_s h}{C_p} g\alpha \, \Delta t & \text{for} \quad Q_n < 0 \\ 0 & \text{for} \quad Q_n \geq 0 \end{cases}$$

and Q_n = net heat across air–water interface
  h = depth.

If the mixed layer depth increases from depth h to $h + dh$ in a time dt, then

$$\frac{dh}{dt} = U_e, \text{ which is the entrainment velocity} \tag{9.11}$$

which may be related to the Richardson number:

$$\frac{U_e}{\sigma} \propto \frac{1}{R_i}$$

The rate of deepening of the mixed layer may also be related to the coefficient of eddy diffusivity, K_H:

$$K_H = h \frac{dh}{dt} \tag{9.12}$$

Typical value for the rate of deepening is

$$\frac{dh}{dt} \simeq 10^{-6} \, \text{m}^2/\text{s}$$

9.2.3 Hydrodynamic Models

Mixing in lakes and reservoirs is brought about by various mechanisms, notably the following:

(a) *Wind.* Although frictional transfer of wind energy is limited to the top 2–3 m of the water column, nevertheless the wind is the principal mechanism causing mixing in most lentic water bodies. In stratified lakes the wind causes the formation of long period waves (seiches) which are responsible for mixing in both the epilimnion and the hypolimnion.

(b) *Inflow/outflow.* In most large lakes and reservoirs the mixing caused by this mechanism is limited to local effects but in water bodies with short retention periods (10 days or less) the exchange of water may cause destratification or other important effects.

For lakes and reservoirs with significant spatial differences in water quality due to wind-induced, or inflow–outflow induced circulations, the usual modelling approach has been a two-dimensional (depth integrated) mass transport, defined by

$$(U, V) = \int_{H}^{S} \frac{\rho \bar{v}}{\rho_0} \, dz \qquad (9.13)$$

where H is the depth from the surface;
   S is the surface displacement;
   ρ is the specific gravity;
   ρ_0 is the specific gravity of the inflow water;
   v is the vector of transport in the U and V directions;
   z is the depth axis.

The mass transport may be related to wind stress and/or bottom stress suitably modified for Coriolis effects and inertial inputs for the inflow. The inflow will find its own density level and so may behave as interflow, underflow and overflow, depending upon the Froude number;

$$F = \frac{Q}{g \, H^5} \qquad (9.14)$$

Such models have been chiefly used for studying flow patterns and sediment transport in recreational lakes, storm water impoundments, river storage lakes and similar bodies of water with relatively short retention times (see Henderson-Sellers (1984) for further details and for examples of applications of this type of model).

However, in the majority of lakes and reservoirs, horizontal spatial patterns of water quality are relatively important and the water body can be treated as a two- or three-component system as shown in Figure 9.5. Such situations may be modelled in purely chemical or chemical and biological terms. These are discussed in Sections 9.3 and 9.4 respectively.

9.3 CHEMICAL MODELS

As explained in Section 9.2, most reservoirs and lakes have retention times which are sufficient to allow horizontal mixing to occur. They may therefore be represented as stirred tanks as shown in Figure 9.4 with the following provisos:

(a)   Where vertical stratification occurs it is necessary to consider the water body as composed of two stirred tanks. As shown in Figure 9.5 this may be extended to a three-compartment system when sediment plays an important role.

(b)   Where lakes and reservoirs are a considerable size (> 50–$100 \, km^2$) it may be necessary to subdivide the total area into a number of stirred tanks in series.

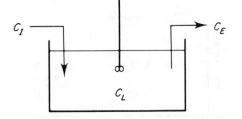

Figure 9.4. Diagrammatic representation of a lake as a stirred tank reactor (*Note*: $C_E = C_L$; $C_L \neq C_1$)

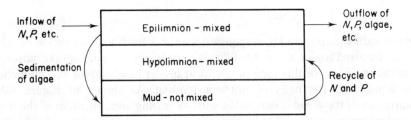

Figure 9.5. Model of a stratified lake and sediment as a three-component system

With each stirred tank the rate of change of a pollutant with first-order decay can be expressed as

$$\frac{dC_2}{dt} = \frac{M}{V} - \frac{Q}{V} \times C_2 - K \times C_2 \tag{9.15}$$

where C_2 = concentration of solute in reservoir
  M = mass loading rate = $Q/V*C_1$
  C_1 = concentration in inflow
  V = volume of reservoir
  K = rate coefficient for decay of solutes.

This equation can then be used to explore the relationship between loading rate and resultant concentration (in the lake and in the outflow). For example Figure 9.6 shows the response of a lake to increased and decreased loading of chloride. Note the considerable length of the response time and note also that the graph can be calculated using the analytical solution

$$C_{2t} = C_{20} + (C_1 - C_{20})e - \left(\frac{Qt}{V}\right) \tag{9.16}$$

where C_{20} = initial concentration in the lake.

This approach has been used to model concentrations of various substances, mainly nutrients and toxins; it is not suitable for the modelling of sediments since the key

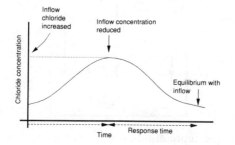

Figure 9.6. Changes in the chloride concentration in a lake in response to increase and decrease in flow concentration

process of sedimentation and re-suspension require a much finer definition of velocity. Even for dissolved toxins it may be difficult to properly simulate uptake, adsorption, etc. So the main use of this type of chemical model has been for the prediction of eutrophication on the basis of nutrient loading. As shown in Figure 9.7, the concentration of the algae is correlated with increasing concentration of the limiting nutrient (generally considered to be phosphorus). Because of the generalized nature of the relationship between algae and phosphorus, which ignores site-specific factors like climate, hydrology etc. the time of onset of algal growths is not precisely forecast. However, the model is attractive because of the minimal data requirement.

The forecast is based upon the relationship between the average concentration of phosphorus in the inflow $\overline{P_i}$, and the resulting concentrations of algae and phosphorus in the lake (A_L and P_L respectively) as a function of the average retention time, T. The relationship may be used to calculate the required level of phosphorus in the inflow to prevent excessive algal growth, as shown in Figure 9.8.

$$\text{Total phosphorus load} = \frac{\overline{P_0}}{T} zA \qquad (9.17)$$

where z = mean depth
A = surface area
T = time

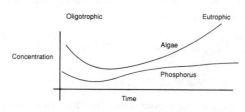

Figure 9.7. Eutrophication model

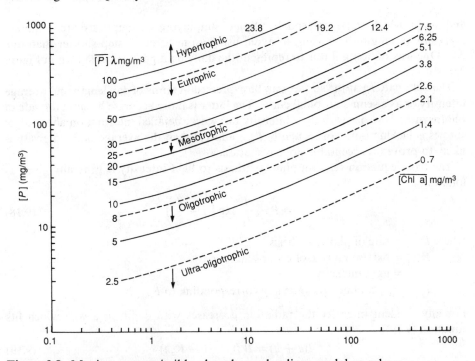

Figure 9.8. Maximum permissible phosphorus loadings on lakes, where
$[\overline{P}]$ is average annual concentration of total phosphorus;
$[\overline{Chl\ a}]$ is average annual concentration of chlorophyll a in the eutrophic zone;
$T(w)$ is water residence time

9.4 BIOLOGICAL MODELS

Most of the important changes in water quality in reservoirs and lakes occur as the result of the activities of microorganisms, especially algae and it is, therefore, rarely possible to satisfactorily simulate these changes by purely physical or chemical models. It is particularly important to find explicit biological models for operational control of reservoirs but biological models have also proved useful as design aids.

The individual processes of algal growth, death, nutrient uptake etc. are discussed separately below, followed by some suggestions for ways of incorporating these processes into eutrophication models.

9.4.1 Algal Growth

The rate of algal growth is dependent on light and nutrients with temperature playing a secondary role. Growth is due to photosynthesis; its calculation is one of the most complex aspects of lake modelling, as it is so closely controlled by the underwater light intensity, which is a function of the depth, colour and turbidity and the length of day and night which varies seasonally. Some simplifying assumptions are required, especially in long-term models, to avoid using an integration step shorter than one day. One way of doing this is to represent the daily light patterns as shown in Figure 9.9.

The light pattern at the surface can be expressed in terms of day length and average intensity or as a semi-sinusoidal curve. The latter is more accurate because the rate of photosynthesis varies with light intensity but where detailed information about algal species is lacking (and consequently P_{max} and I_k) then the average daily intensity is likely to prove an acceptable approximation.

The basic equation relation photosynthesis to light intensity is generally given as follows:

$$P = P_{max} \frac{I}{I_k} \exp\left(1 - \frac{I}{I_k}\right) \tag{9.18}$$

where P = rate of photosynthesis
P_{max} = maximum rate of photosynthesis
I = light intensity
I_k = optimum light intensity corresponding to P_{max}

For any incident intensity the radiation decreases with depth in a way which fits equation (9.19):

$$I(z + h) = I(z) \exp(-K_e h) \tag{9.19}$$

$I(z)$ and $I(z + h)$ are light intensities at depth (z) and $(z + h)$ respectively.

It is possible to put all these factors together and obtain an expression from the daily photosynthesis in a water column integrated over depth. This is given by

$$GP = \frac{NP_{max}}{K_e} 0.6\left(1.33 \sin h^{-1}\phi - \frac{1}{\phi}(\sqrt{(1 + \phi^2)} - 1)\right) \tag{9.20}$$

where GP = gross daily production per unit area = day length
N = algal concentration

$$\phi = \frac{I}{I_k}$$

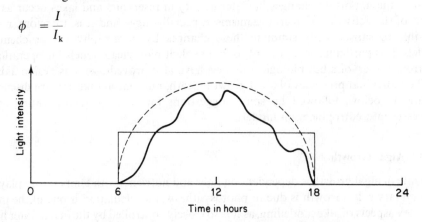

Figure 9.9 Methods of simulating the daily pattern of solar radiation

The rate of photosynthesis increases with temperature up to an optimum and thereafter decreases. This may be expressed in a similar way to the light response shown in equation (9.21):

$$P_t = P_{max} \frac{t}{t_{opt}} \exp\left(1 - \frac{t}{t_{opt}}\right) \tag{9.21}$$

where P_t = rate of photosynthesis at temperature t
t_{opt} = optimum temperature giving P_{max}

In many reservoirs and lakes the above equations are not adequate for predicting the growth of algal populations because of the limitations imposed by lack of nutrients. This additional feature can be incorporated by using a Michaelis–Menten equation for the limiting nutrients

$$P_L = P\left(\frac{N}{K_s + N}\right) \tag{9.22}$$

where P_L = rate of growth with nutrients limitation
P = rate of growth in absence of nutrients limitation
N = concentration of limiting nutrients.

But this approach fails to take account of the ability of algae to store nutrient and equation (9.22) therefore tends to predict prematurely the end of an algal bloom. A more satisfactory simulation can be obtained from a two-compartment model in which nutrients are taken up at a rate given in equation (9.23):

$$U = U_{max} \frac{Q_m - Q}{Q_m - K_Q}\left(\frac{N}{K_s + N}\right) \tag{9.23}$$

where U and U_{max} are the actual and maximum uptake rates of nutrients
Q and Q_m are the quantity stored in algae and the maximum storage

and the growth rate is determined by the stored nutrient according to equation (9.24):

$$G = G_{max}\left(\frac{Q - K_Q}{Q}\right) \tag{9.24}$$

where G is the growth rate
K_Q is a constant representing the minimum value of Q in the cell.

9.4.2 Respiration

In algal cells a significant proportion of the organic material produced by photosynthesis is used in respiration. Fortunately the rate of respiration may be represented very simply as

$$\frac{dN}{dt} = rN \tag{9.25}$$

where r = rate of respiration
N = algal concentration.

This equation needs a temperature correction when large variations in temperature occur. The correction is similar to that used for photosynthesis.

9.4.3 Sedimentation

Some algal cells appear to be able to control their buoyancy but some, especially diatoms, have a tendency to sink and all species cease to be buoyant when dead. Sedimentation is therefore an important process in removing algae.

There is a complication in representing sedimentation because where the algae are still alive resuspension and re-growth is possible. The hydraulics are also complicated since the shape factor for algae is not easily defined and is constantly changing due to small-scale turbulence.

The simplest basis to represent sedimentation losses is a constant portion of the standing crop:

$$S = K_{sed}N \tag{9.26}$$

where S = rate of sedimentation
K_{sed} = sedimentation coefficient.

9.4.4 Predation

In many reservoirs and lakes the effects of predation may be the major factor controlling the population levels of algae. Preferential grazing of some species may occur but in longer-term models the composition of algae and predators is likely to change, so it is best to make the representation very simple

$$\frac{dN}{dt} = -gNZ \tag{9.27}$$

where g = filtering rate of predators
Z = predator concentration.

The use of equation (9.27) obviously requires a prediction of the predator population. The kinetics of the predators can be based on the concept of yield and the rate of utilization of phytoplankton. This therefore becomes

$$\frac{dZ}{dt} = \frac{-gNZ}{Yz} \tag{9.28}$$

where Yz = yield of predators per unit mass of algae consumed.

An alternative approach that has been used is to regard the growth of zooplankton in terms of Michaelis–Menten kinetics with the algae as the substrate. The equation for growth therefore becomes

$$\frac{dZ_j}{dt} = \left[\sum_{i=1}^{m} u_{i,j} Y_{i,g} B_j \left(\frac{N_i}{B_j + N_i} \right) \right] g_j Z_j \tag{9.29}$$

where $\mu_{i,j}$ = growth rate of species Z_i on alga N_j
    Y_i = yield coefficient
    g = predation coefficient for species Z
    B = Michaelis–Menten coefficient

The predation of higher tropic levels may be represented using similar equations but the complexities increase due to variation in fecundity and survival at different ages (see Chapter 5.6).

9.4.5 Nutrient Recycle

This occurs as a result of the bacterial decomposition of algal cells in the bottom mud. The rate of regeneration is therefore primarily dependent upon the concentration of organic matter in the mud. Since this is almost solely derived from the sedimented algae, the rate of regeneration can be simply represented as some function of the rate of sedimentation:

$$R = f(s) \tag{9.30}$$

where s = rate of sedimentation.

When the limiting nutrient is phosphorus, the value of f is in two ranges—very low or very high, depending whether the mud is aerobic or anaerobic. Under aerobic conditions most of the regenerated phosphorus (90–99%) is locked up in ferric iron complexes. If the overlying water becomes anaerobic, then the complexes break down and the value of f approximates to unity. It is therefore necessary to carry out an oxygen balance on the overlying water (see Section 9.4.6).

9.4.6 Oxygen Balance

In the case where a lake or reservoir stratifies, the oxygen balance is on the hypolimnion; this is effectively cut-off from re-aeration, so the balance is merely

$$DO_t = DO_0 - \frac{RK_D SA}{V_H} \tag{9.31}$$

where DO_0 and DO_t are the oxygen concentrations at the beginning and at any time t
    K_D converts the rate of benthal bacterial activity into oxygen units
    SA and V_H are the surface area of mud and the volume of hypolimnion.

When the water body is not stratified, the oxygen balance is much more complex since other processes like photosynthesis, respiration, re-aeration and inflow and outflow are important. Fortunately, in these circumstances it is very rare for the lake water to become so deoxygenated as to cause liberation of phosphorus from ferric complexes.

9.5 LAKE MODELS

Lake models may be classified into the following categories:

(a) Long-term planning models whose aim is to forecast the increase in algal growth in a reservoir or to predict the effect of some management decision, e.g. water transfer, watershed management, etc.

(b) Short-term operational models which are concerned with the day-to-day control of algal growth in an already eutrophic reservoir by induced circulation, algicides, etc.

Although the basic structure of these two types of model is similar, there are sufficient differences in detail to make it worth while discussing them separately.

9.5.1 Long-term Planning Models

The long-term changes in the productivity of a lake or reservoir may be represented diagrammatically as shown in Figure 9.10.

Most models are concerned with the period of increasing productivity as the lake passes from oligotrophy to eutrophy and aim at forecasting the rate of change. The models vary in the hydraulics complexity and may be classified as

(a) zero-dimensional—whole water body regarded as a stirred tank reactor;

(b) compartment model—water body divided into two horizontal layers and may also be divided into segments where lateral heterogeneity occurs;

(c) one-dimensional—usually the dimension is vertical and allows for concentration gradients, light intensity gradients, etc.;

(d) two- and three-dimensional—usually cater for vertical variations with longitudinal and or lateral variations where these are important.

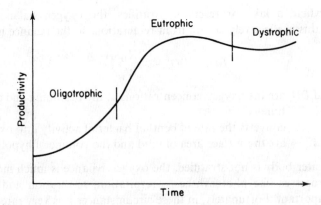

Figure 9.10. Long-term changes in the productivity of a lake

The choice depends upon the size of the impoundment, the precision required and the data available. It should be remembered that an increase in one dimension increases the data requirements by an order of magnitude.

The following example is a compartment model in which the water body is divided into epilimnion and the hypolimnion during the summer period and is fully mixed during the winter. The length of summer and winter are arbitrarily fixed at 165 days and 200 days respectively. The transition between stratified and unstratified conditions is regarded as instantaneous. Starting from the beginning of summer, the equations may be written in the form of daily balances on algae and nutrients in the epilimnion and dissolved oxygen and nutrients in the hypolimnion, as follows.

(1) *Epilimnion*:

$$\frac{dN}{dT} = GP - R - Sed - \frac{Q_2 * N}{V_E} \qquad \text{algal balance} \qquad (9.32)$$

where

algal production per day $\quad GP = \dfrac{NP_{max}\Delta}{K_e} 0.6 \left[1.33 \, (\sinh^{-1} \phi) \dfrac{1}{\phi} (\sqrt{(1 + \phi^2)} - 1) \right]$

$$\phi = \frac{I}{I_k}$$

The light intensity is usually given as an average daily intensity for each month.

$$R = - rN \qquad \text{(respiration per day)} \qquad (9.33)$$

$$Sed = K_{sed}N \qquad \text{(sedimentation per day)} \qquad (9.34)$$

$$Q_2 = \text{daily outflow and } V_E = \text{volume of epilimnion} \qquad (9.35)$$

$$\frac{dC_2}{dt} = \frac{Q_1}{V_E} C_1 - \frac{Q_2}{V_E} C_2 - \mu \qquad \text{nutrient balance} \qquad (9.36)$$

where

$$\frac{Q_1}{V_E} C_1 = \text{daily influx of nutrient}$$

$$\frac{Q_2}{V_E} C_2 = \text{daily efflux of nutrient}$$

and

$$u = u_{max} \left(\frac{Q_m - Q}{Q_m - K_Q} \right) \left(\frac{C_2}{K_s - C_2} \right) \qquad \text{uptake rate per day}$$

where Q_m and Q are the maximum and current amounts of nutrient stored and K_Q is the Michaelis–Menten coefficient

(2) *Hypolimnion*

$$\frac{dC_3}{dt} = \frac{\mathscr{F}(\text{sed})}{V_4} \tag{9.37}$$

where C_3 = concentration of nutrients in hypolimnion
V_4 = volume of hypolimnion

$$\frac{dDO_H}{dt} = \frac{RK_DSA}{V_H} \tag{9.38}$$

Each day during the summer period the mass balance is calculated and the concentrations of algae, nutrients and oxygen are adjusted. At the end of the summer period the water body becomes fully mixed and the resulting concentrations of nutrients and algae become

$$N_L = \frac{N_E V_E + N_H V_H}{V_E + V_H} \tag{9.39}$$

$$C = \frac{C_E V_E + C_H V_H}{V_E + V_H} \tag{9.40}$$

These are the starting concentrations for the winter period, during which the daily balances are given by

$$\frac{dN_L}{dt} = GP - R - \text{sed} - \frac{Q}{V_L} N \tag{9.41}$$

$$\frac{dC_2}{dt} = \frac{Q_1}{V_L} C_1 - \frac{Q_2}{V_L} C_2 - u + \mathscr{F}(\text{sed})$$

A dissolved oxygen balance is not normally required. At the end of the winter period the lake is divided arbitrarily into epilimnion and the hypolimnion and the cycle begins again.

The flow diagram for this type of model is shown in Figure 9.11 and a listing of the program is given in the Appendix to the chapter.

9.5.2 Short-term Operational Models

The two major differences between short-term and long-term models are as follows:

(a) In short-term models it is possible to represent the kinetics of the dominant species and to give similar detailed treatment to other inputs like solar radiation, temperature, nutrient inflow, etc.

(b) In short-term models it is possible to take account of predation and to take account for succession in the dominant species of prey and predator which will vary with time.

Because short-term models produced forecasts for weeks ahead rather than years, it is possible to reduce the size of the integration step below a day, which reduces the numerical errors.

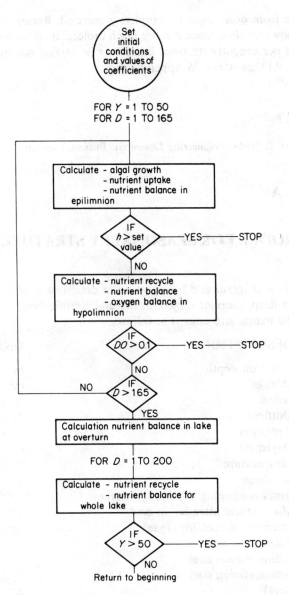

Figure 9.11. Flow diagram for lake model

9.5.3 Tropical Lakes and Reservoirs

There is a tendency for deep lakes and reservoirs in tropical areas to stratify permanently. In these circumstances the water quality in the hypolimnion becomes significantly different from the epilimnion such that chemical stratification is superimposed upon thermal stratification. The hypolimnion is permanently anaerobic so that nutrient release from dead algae is completely recycled. Return of nutrients to the epilimnion is however slow, since it relies upon molecular diffusion.

Modelling of permanently stratified lakes may be carried out by suitable amendment to Figure 9.11, as shown in Appendix 2.

REFERENCE

Henderson-Sellers, B. (1984) *Engineering Limnology*. Pitman, London.

APPENDIX A

1 LAKE MODEL FOR SEASONALLY STRATIFIED LAKES

The following flow diagram and listing give the details of a lake model which was formulated for a deep reservoir subject to thermal stratification.

The key to the names and units is as follows:

NAME	DESCRIPTION	UNITS
DEPTH	Reservoir depth	m
STOR	Storage	m³
IN	Inflow	m³/day
OUT	Outflow	m³/day
RAD	Radiation	cal/(m² day)
DAYL	Daylength	h
TEMP	Temperature	°C
SUN	Sunshine	h
VOLP	Array containing depths, % volumes	
KMIN	Min. vertical attenuation coef.	/m
K1	Attenuation coef. for algae	/(m μg Chla.)
R	Respiration coef.	
S1	Sedimentation coef.	
S2	Sedimentation coef.	
Z	Depth	m
VP	% Volume	
N	Conc. of chlorophyll	μg Chla./l
NUTU	Nutrient uptake	mg/(μg Chla.)
NUTL	Coef. of nutrient limitation	
PHI	Ratio of IOMAX:IK	
F	Light function	
GP	Gross photosynthesis	μg Chla./m²
RESP	Respiration losses	μg Chla./m²

SED	Sedimentation losses	μg Chla./m^2
SEDAV	Average sedimentation	μg Chla./m^2
PROD	Net production of chlorophyll A	μg Chla./m^2
NL	Nutrient limitation function	
NU	Nutrient uptake function	
ZAV	Average depth of reservoir	m
ASTOR	Average storage	m^3
PMAX	Max. rate of photosynthesis	mg O$_2$/(μg Chla. h)
IK	Light intensity	kerg/(cm^2 s)
IOMAX	Max. light intensity	kerg/(cm^2 s)
LIGHT	Light function	

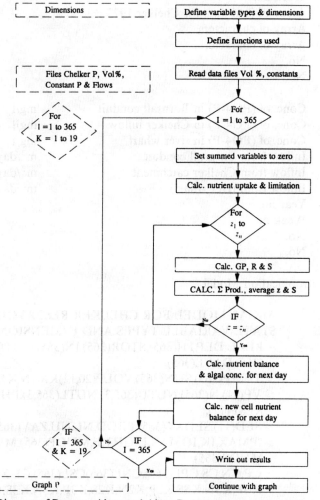

Diagram of Program Algae and Algae P (differences shown in dotted).

E	Effective light attentuation coef.	/m
MUDP	Concentration of (PO4-P) in the mud	g/m^2
CP	Concentration of (PO4-P) in inflow	mg/l
CPL	Concentration of (PO4-P) in res.	mg/l
X	Min. nutrient limitation coef.	
KSP	Michaelis–Menten constant	mg/l
K3P	Rate of return of (PO4-P) from mud	
UMP	Max. uptake of (PO4-P)	mg/(μg Chla. day)
CPC	Concentration of (PO-P) in algae	mg/(μg Chla. day)
KQP	Min. concentration of (PO4-P) in algae	mg/(μg Chla. day)
QMP	Max. concentration of (PO4-P) in algae	mg/(μg Chla. day)
DAY	Day no.	
Y	Week no.	
FIELD	Array of field data for Chelker 1977	
YLINE	Array of characters	
Line	Array of characters	
A	No.	
Q	No.	
CPB	Conc. of (PO4-P) in Burnsall conduit	mg/l
CPCH	Conc. of (PO4-P) in Chelker inflow	mg/l
CPW	Conc. of (PO4-P) in river wharf	mg/l
INB	Inflow from Burnsall conduit	m^3/day
INCH	Inflow from Chelker catchment	m^3/day
INW	Inflow from river wharf	m^3/day
YEAR	Year no.	
B	Weak no.	
C	No.	
Y	No.	

Listing of Eutrophication Model

```
1    C     ALGAL MODEL FOR CHELKER RESERVOIR
2    C     STATE VARIABLE TYPES AND DIMENSIONS
3          REAL DEPTH(365),STOR(365),IN(365),OUT(365),RAD
             (365),DAYL(365),
4          > TEMP(365),SUN(365),VOLP(20,12)KMIN,K1,2,S1,S2,Z(11),
5          > VP(11),N(365),NUTU(365,3),NUTL(365,3),PHI(2),F,GP(11),
             RESP(11),
6          > SED(11),SEDAV(365),PROD,NL,NU,ZAV(365),ASTOR
7          > ,PMAX,IK,IOMAX,LIGHT,E,MUDP(365),MUDN(365),
             MUDS(365),
8          > CP,CN,CS,CPL(365),CNL(365)CSL(365),CL,X(365),
9          > KSP,KSN,KSS,K2P,K2N,K2S,K3P,K3N,K3S,UMP,
10         > UMN,UMS,CPC(365),CNC(365),CSC(365),
```

```
11              > KQP,KQN,KQS,QMP,QMN,QMS
12                INTERGER DAY(365),Y
13       C        DEFINE FUNCTIONS
14                PMAX(TEMP) = 3.1*EXP(0.09*TEMP)
15                IK(TEMP = (PMAX(TEMP))/0.63
16                IOMAX(RAD,DAYL,SUN) = 10.708*RAD*0.96*(0.18 +
                  (0.55*SUN/DAYL))/DAYL
17                LIGHT(PHI) = (1.333*ATAN(0.5*PHI)) -
                  ((ALOG(1 + (0.5*PHI)**2.0))/PHI)
18                E(N) = (KMIN + K1*N)*1.25
19                NL(CQ,CKQ) = (CQ = CKQ)/CQ
20                NU(C,CQ,CQM,CKQ,CKS) = ((CQM - CQ)/
                  (CQM - CKQ))*(C/(CKS + C))
24       C        READ IN INPUT DATA(FILES CHELKER77,VOL%,
                  CONSTANTS)
25                DO 10 I = 1,365
26                READ(5,20)DAY(I),DEPTH(I),STOR(I),IN(I),OUT(I),
                  RAD(I),DAYL(I),
27                1 TEMP(I),SUN(I)
28       10       CONTINUE
29       20       FORMAT(13,F6.2,F6.3,2F6.2,F6.1,F6.2,2F5.1)
30                DO 30 I = 1,20
31                READ(5.40)(VOLP(I,J),J = 1,12)
32       30       CONTINUE
33       40       FORMAT(12F6.2)
34                READ(5,50)KMIN,K1,R,S1,S2,AC,CP,CN,CS
35                READ(5,51)KSP,KSN,KSS,UMP,UMN,UMS,
                  K3P,K3N,K3S
36                READ(5,52)KQP,KQN,KQS,QMP,QMN,QMS
37                READ(5,53)N(1),CNL(1),CSL(1),MUDP(1),MUDN(1),
                  MUDS(1)
38                READ(5.54)CPC(1),CNC(1),CSC(1)
39       50       FORMAT(F5.2,F6.3,3F5.2,F6.3,F7.4,2F7.3)
40       51       FORMAT(3F6.3,1P3E9.2,OP3F5.2)
41       52       FORMAT(1P6E9.2)
42       53       FORMAT(F6.3,F7.4,2F6.2,1P3E9.2)
43       54       FORMAT(1P3E9.2)
44       C        BEGIN DAILY CALCULATIONS
45                DO 60 I = 1,365
46       C        DETERMINE DEPTHS FOR % VOLUME
47                DO 70 J = 1,19,2
48                IF(ABS(VOLP(J,1) - DEPTH(I)).LE.0.5) GOTO 80
49       70       CONTINUE
50       75       FORMAT('DEPTH(',13,') OUTSIDE RANGE OF
                  VOLP')
```

```
52        C       SET INITIAL CONDITIONS BEFORE SUMMING OVER
                  RANGE OF Z
53       80       PROD = 0.0
53.2              DELTA = 0.0
54                ZAV(I) = 0.0
55                SEDAV(I) = 0.0
56                NUTU(I,1) = NU(CPL(I),CPC(I),QMP,KQP,KSP)*UMP
57                NUTU(I,2) = NU(CNL(I),CNC(I),QMN,KQN,KSN)*UMN
58                NUTU(I,3) = NU(CSL(I),CSC(I),QMS,KQS,KSS)*UMS
59                NUTL(I,1) = NL(CPC(I),KQP)
60                NUTL(I,2) = NL(CNC(I),KQN)
61                NUTL(I,3) = NL(CSC(I),KQS)
62                X(I) = NUTL(I,1)
63                IF(NUTL(I,2).LT.X(I))X(I) = NUTL(I,2)
64                IF(NUTL(I,3).LT.X(I))X(I) = NUTL(I,3)
65                PHI(1) = IOMAX(RAD(I),DAYL(I),SUN(I))/IK(TEMP(I))
66                DO 90 K = 2,12
67                Z(K − 1) + (VOLP(J,K) + DEPTH(I) − VOLP(J,1))*0.3048
68                VP(K − 1) = VOLP(J + 1,K)
69        C       DETERMINE GP,R,&S FOR EACH Z
70                PHI(2) = (IOMAX(RAD(I),DAYL(I),SUN(I))*EXP( − E
                  (N(I))*Z(K − 1)))
70.5             > /IK(TEMP(I))
72                F = LIGHT(PHI(1)) − LIGHT(PHI(2))
73                GP(K − 1) = 0.375*AC*N(I)*PMAX(TEMP(I))*DAYL(I)*
                  1.2*F*X(I)/E(N(I))
74                RESP(K − 1) = 0.375*AC*24*R*PMAX(TEMP(I))*
                  Z(K − 1)*N(I)
75        C       SEDIMENTATION LOSS DEPENDENT ON STRESS
76                IF(X(I).GE.0.6)SED(K − 1) = S1*N(I)*Z(K − 1)
77                IF(X(I).LT.0.6)SED(K − 1) = S2*N(I)*Z(K − 1)
78        C       CALCULATE PRODUCTION FOR EACH DEPTH & SUM
79                IF(Z(K − 1).GT.0.0)
80               > PROD = PROC + ((GP(K − 1) − RESP(K − 1) −
                  SED(K − 1))*VP(K − 1)/Z(K − 1))
80.2              IF(Z(K − 1).GT.O.O)
80.4             > DELTA = DELTA + (GP(K − 1) − RESP(K − 1))
                  *VP(K − 1)/Z(K − 1)
81        C       CALCULATE AVERAGE SEDIMENTATION & DEPTH
                  FOR WHOLE RESERVOIR
82                SEDAV(I) = SEDAV(I) + SED(K − 1)*VP(K − 1)
83                ZAV(I) = ZAV(I) + Z(K − 1)*VP(K − 1)
84       90       CONTINUE
85        C       CALCULATE NEW ALGAL CONCENTRATION
86                IF(I.L.T.365)
```

```
87                  > N(I + 1) = PROD + (N(I)*(STOR(I)*1000.0 − OUT(I))/
                      (STOR(I + 1)*1000.0))
88                    IF(N(I + 1).LT.0.000001)N(I + 1) = 0.000001
89        C           CALCULATE NUTRIENT BALANCE FOR NEXT DAY
90                    IF(I.EQ.365) GOTO 100
91                    ASTOR = (STOR(I) + STOR(I + 1))*1000.0/2.0
91.5      100         IF(I.GT.1)
93                  > MUDP(I) = 0.8*SEDAV(I − 1)*CPC(I − 1) +
                      MUDP(I − 1)*(1 − K3P)
93.2                  IF(I.GT.1)
94                  > MUDN(I) = 0.8*SEDAV(I − 1)*CNC(I − 1) +
                      MUDS(I − 1)*(1 − K3S)
94.2                  IF(I.GT.1)
95                  > MUDS(I) = 0.8*SEDAV(I − 1)*CSC(I − 1) +
                      MUDS(I − 1)*(1 − K3S)
96                    IF(I.EQ.365) GOTO 60
96.4                  CPL(I + 1) = CPL(I) + (IN(I)*CP/ASTOR) − (OUT(I)*
                      CPL(I)/ASTOR
97                  >  + (K3P*MUDP(I)/(ZAV(I)))-(NUTU(I,1)*N(I))
99                    CNL(I + 1) = CNL(I) + (IN(I)*CN/ASTOR) −
                      (OUT(I)*CNL(I)/ASTOR)
99.5                >  + (K3N*MUDN(I)/(ZAV(I))) − (NUTU(I,2)*N(I))
101                   CSL(I + 1) = CSL(I) + (IN(I)*CS/ASTOR) −
                      (OUT(I)*CSL(I)/ASTOR)
101.2               >  + (K3S*MUDS(I)/(ZAV(I))) − (NUTU(I,3)*N(I))
101.4                 IF(CPL(I + 1).LT.0.0)CPL(I + 1) = 0.0
101.6                 IF(CNL(I + 1).LT.0.0)CNL(I + 1) = 0.0
101.8                 IF(CSL(I + 1).LT.0.0)CSL(I + 1) = 0.0
103       C           CALCULATE NEW CELL NUTRIENT CONCENTRA-
                      TIONS
104                   CPC(I + 1) = (CPC(I) + NUTU(I,1))*N(I)/(N(I) + DELTA)
105                   IF(CPC(I + 1).GE.QMP)CPC(I + 1) = QMP
106                   IF(CPC(I + 1).LE.KQP)CPC(I + 1) = KQP
107                   CNC(I + 1) = (CNC(I) + NUTU(I,2))*N(I)/(N(I) +
                      DELTA)
108                   IF(CNC(I + 1).GE.GMN)CNC(I + 1) = GMN
109                   IF(CNC(I + 1).LE.KQN)CNC(I + 1) = KQN
110                   CSC(I + 1)(CSC(I) + NUTU(I,3))*N(I)/(N(I) + DELTA)
111                   IF(CSC(I + 1).GE.GMS)CSC(I + 1) = QMS
112                   IF(CSC(I + 1).LE.KQS)CSC(I + 1) = KQS
113       60          CONTINUE
114       C           WRITE OUT RESULTS
115                   WRITE(6,119)
116                   WRITE(6,110)
117                   WRITE(6,111)
```

```
118      110      FORMAT('KMIN K1 2 S1 S2 AC CP CN CS')
119      111      FORMAT('/M /M.A FRAC FRAC A:C MG/L MG/L
                  MG/L')
120               WRITE(6,50)KMIN,K1,2,S1,S2,AC,CP,CN,CS
121               WRITE (6,119)
122      119      FORMAT('...........................................................................')
123               WRITE(6,120)
124               WRITE(6,121)
125      120      FORMAT('WEEK ALGA PO4 − P NO3 − N SI02
                  LNUT NL M
125.2            > UDP MUDN MUDS')
126      121      FORMAT('(NO) (UGA/L) (MG/L) (MG/L) (MG/L) (G/S)
126.2            > QM) (G/SQM) (G/SQM)')
127               DO 130 I = 1,365,7
127.01           Y = (DAY(I) + 6)/7
127.05           W = CSL(I)*15/7
128               IF(X(I).EQ.NUTL(I,1))
129              > WRITE(6,140)Y,N(I),CPL(I),CNL(I),W,
                   X(I),(MUDP(I),MUDN(I),MUDS(I)
130               IF(X(I).EQ.NUTL(I,2))
131              > WRITE(6,141)Y,N(I),CPL(I),CNL(I),W,X(I),MUDP(I),
                   MUDN(I),MUDS(I)
132               IF(X(I).EQ.NURL(I,3))
133              > WRITE(6,142)Y,N(I),CPL(I),CNL(I),W,X(I),MUDP(I),
                   UMDN(I),MUDS(I)
134      130      CONTINUE
135      140      FORMAT(15,1PE8.1,1P3E10.3,'PO4 − P',OPF4.1,1P3E9.2)
135.6    141      FORMAT(15,1PE8.1,1P3E10.3,'NO3 − N',OPF4.1,1P3E9.2)
136.2    142      FORMAT(15,1PE8.1,1P3E10.3,'SIO2',OPF4.1,1P3E9.2)
138               &CONTINUE WITH GRAPH RETURN
184               END
```

2 MODELS FOR LAKES WITH OTHER THERMAL PATTERNS

The program described above may be modified to accommodate other patterns of stratification. For example, if the lake is permanently stratified, the following modifications would apply:

(i)   Hypolimnion can be assumed to be sufficiently anaerobic for complete nutrient recycle to occur.

(ii) Transfer of nutrients between hypolimnion and epilimnion will take place mainly by molecular diffusion

(iii) Inflow and outflow will mix only with epilimnion.

(iv) There will be very limited seasonal changes in lake temperature and algal growth.

When no stratification occurs, or it occurs only briefly, then the program may be modified by assuming

(i) complete mixing vertically;
(ii) aerobic conditions with slow recycle of nutrients together with appropriate seasonal patterns for temperature and algal growth.

_____Chapter 10

Groundwater Quality Modelling

R. MACKAY AND M. S. RILEY

10.1 INTRODUCTION

Contamination of groundwater, both past and present, is a major problem. The adoption of intensive agricultural practices has led to high levels of nitrates as well as pesticides and herbicides being observed in groundwater. Increasingly, industrial effluents leached from waste tips or from accidental spillages are being detected in groundwater abstractions. Urban effluents from road drainage and leaking sewer systems have also contributed noticeably to the deterioration of the quality of underlying groundwaters. In many countries steps are being taken to ensure that contamination from each of these sources is reduced. Unfortunately, the damage that has already occurred cannot be easily undone. Unlike surface waters, groundwaters tend to move very slowly. Residence times for groundwater in an aquifer may be of the order of tens or hundreds of years. As a consequence, contamination of an aquifer system can go undetected over long periods of time and lead to extensive degradation of the quality of the groundwater resource.

Once groundwater contamination has been detected, the resource manager is faced with identifying the following:

(1) the source of the contamination;
(2) the spatial extent of the contamination;
(3) the future impact of the contamination on groundwater abstractions and surface water systems;
(4) the range of possible strategies for mitigating the impact of the contamination and their associated costs.

A model describing the contaminant transport characteristics of the groundwater regime and the chemical interactions of the particular contaminant(s) of interest can

---

An Introduction to Water Quality Modelling. 2nd Edition. Edited by A. James

provide a powerful tool for solving each of these problems. There are four main stages for the development of such a model, the development of a conceptual model, the expression of the conceptual model in the form of a mathematical model, the numerical or analytical solution of the mathematical model and, finally, the validation of the model using the available data.

The development of the basic mathematical and numerical aspects of groundwater quality modelling form the primary focus of this chapter. However, the fundamental importance of the conceptual modelling stage to the development of the final quality model cannot be over-emphasized. The conceptual model comprises the set of simplifying assumptions used to define the mathematical model. Thus, the conceptual model is the main link between the mathematical model and the real aquifer system. Much of any groundwater resource is hidden from view. The structure of the aquifer formations, the hydraulic properties of the aquifer system and the distribution of the groundwater flow patterns within it often have to be inferred from sparsely distributed data. Only through very careful development of the conceptual model can the groundwater 'jig-saw' be satisfactorily completed.

Whilst many of the terms used in this chapter are defined at the point where they first appear, it has been assumed generally that the reader has some familiarity with the basic concepts of groundwater occurrence and its exploitation. Freeze and Cherry (1979) provides a comprehensive introduction to groundwater for readers who wish to extend their knowledge of this subject.

10.2 GROUNDWATER FLOW

The understanding of contaminant transport in any aquifer is inextricably linked to the identification of the underlying groundwater flow patterns. In this section the basic equations describing groundwater flow are presented.

10.2.1 The Governing Equation of Groundwater Motion

The flux of groundwater within an aquifer is given by the general form of Darcy's law:

$$q_i = -\frac{k_{ij}}{\mu}\left(\frac{\partial P}{\partial x_j} - \rho g_i\right) \tag{10.1}$$

where q_i is the ith component of the specific discharge, $\mathbf{q}$, $[LT^{-1}]$
     k_{ij} is the ijth component of the intrinsic permeability tensor, $\mathbf{k}$, $[L^2]$
     μ is the fluid viscosity, $[ML^{-1}T^{-1}]$
     P is the fluid pressure, $[ML^{-1}T^{-2}]$
     ρ is the fluid density, $[ML^{-3}]$
and g_i is the ith component of the acceleration due to gravity $(0, 0, -g)$. $[LT^{-2}]$
(*Note*: In this equation and in all subsequent equations, the coordinate axes are aligned such that the x_3 axis points vertically upwards. Additionally, Einstein's summation convention has been adopted throughout this chapter. The subscripts, $i = 1, 2$ and 3 refer to the x, y and z directions respectively.)

In cases where the density of groundwater can be assumed constant, Darcy's law can be reduced to

$$q_i = - K_{ij} \frac{\partial \phi}{\partial x_i} \tag{10.2}$$

where $K_{ij} = k_{ij} \rho g / \mu$ is the ijth component of the hydraulic conductivity tensor, $\mathbf{K}$, $[LT^{-1}]$

and $\phi = P/\rho g + z$ is the hydraulic head. $[L]$

It is important to note that the variables appearing in Darcy's law are all defined as macroscopic averages. Darcy's law describes the average behaviour of groundwater movement within a volume of rock that is much larger than the volume of the individual interconnected voids that transmit the groundwater. This concept of averaging underpins almost all of the development of the mathematics describing both groundwater flow and transport that will be discussed here.

The concept of averaging has been formalized by Bear (1972), among others, who adopts the notion of a representative elementary volume (REV). The magnitudes of the properties, such as hydraulic conductivity, at any point within the aquifer are obtained as averages over an REV centred on the point. The size of the REV is chosen such that any small change in its size would produce an insignificant change in the magnitudes of the aquifer properties (Figure 10.1). The adoption of this approach to defining property values contains two important results. First, the spatial distributions of the derived aquifer properties vary smoothly over the aquifer domain and, therefore, can be interpolated from point observations. Second, the averaged properties can be used in continuum representations of the groundwater flow processes, such as Darcy's law. Both results are essential for the development of usable models. In general the existence of the REV is assumed rather than confirmed during the development of any model.

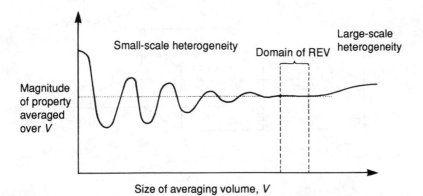

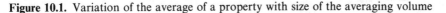

Figure 10.1. Variation of the average of a property with size of the averaging volume

In a volume of rock, the stored mass of groundwater changes in response to changes in fluid pressure. The rate of change of groundwater mass with change of pressure is given by the storage coefficient, S_p $[L^{-2}T^2]$. That is

$$S_p = \frac{\partial M}{\partial P} \tag{10.3}$$

where M is the mass of groundwater per unit volume of rock. $\qquad$ $[ML^{-3}]$

An alternative coefficient is introduced when groundwater density is assumed constant. This is the *specific storativity* $[L^{-1}]$ given by

$$S_s = \frac{1}{\rho} \frac{\partial M}{\partial \phi} \tag{10.4}$$

which represents the change in the stored volume of groundwater per unit volume of rock per unit increase in the hydraulic head. Generally, both storage coefficients are assumed to be constant for problems of practical interest.

The groundwater mass balance at a point, p, in an aquifer is defined using the control volume approach (Figure 10.2). The vector $\mathbf{J}$ $[ML^{-2}T^{-1}]$ with components J_x, J_y and J_z denotes the mass flux (i.e. the mass of groundwater crossing a unit area of aquifer in a unit time). If δM is the increase in M in time δt, then the net increase in the mass of water in the control volume in time δt is $\delta M\, \delta x\, \delta y\, \delta z$. Consideration of the mass flux across each of the six faces of the control volume gives

$$\delta M\, \delta x\, \delta y\, \delta z = \delta t \left\{ \frac{J_x|_{x-\delta x/2,y,z} - J_x|_{x+\delta x/2,y,z}}{\delta x} + \frac{J_y|_{x,y-\delta y/2,z} - J_y|_{x,y+\delta y/2,z}}{\delta y} \right.$$

$$\left. + \frac{J_z|_{x,y,z-\delta z/2} - J_z|_{x,y,z+\delta z/2}}{\delta z} \right\} \delta x\, \delta y\, \delta z \tag{10.5}$$

where $J_{x|x,y,z}$ is the mass flux evaluated at the point (x, y, z).

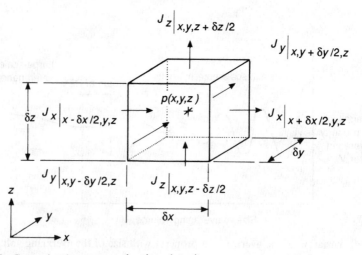

Figure 10.2. Control volume around point $p(x,y,z)$

Dividing this equation by δt, δx, δy and δz, and taking the limit as δt, δx, δy and δz tend to zero gives

$$\frac{\partial M}{\partial t} = -\left\{\frac{\partial J_x}{\partial x} + \frac{\partial J_y}{\partial y} + \frac{\partial J_z}{\partial z}\right\} \tag{10.6}$$

which can be re-written in the form

$$\frac{\partial M}{\partial t} = -\frac{\partial J_i}{\partial x_i} \tag{10.7}$$

Equation (10.7) states that the rate of change of mass per unit volume of aquifer is equal to minus the divergence of the mass flux. This is an important general result which will be used in the development of the mathematical model for contaminant transport.

The water mass balance equation can now be obtained either in terms of the water pressure P or the hydraulic potential ϕ. From equation (10.3) it can be seen that

$$\frac{\partial M}{\partial t} = S_p \frac{\partial P}{\partial t} \tag{10.8}$$

and since the mass flux $\mathbf{J} = \rho\mathbf{q}$, equations (10.1), (10.7) and (10.8) can be combined to give

$$S_p \frac{\partial P}{\partial t} = \frac{\partial}{\partial x_i}\left\{\rho \frac{k_{ij}}{\mu}\left(\frac{\partial P}{\partial x_j} - \rho g_i\right)\right\} \tag{10.9}$$

Similar considerations lead to the mass balance equation expressed in terms of the hydraulic potential, ϕ

$$S_s \frac{\partial \phi}{\partial t} = \frac{\partial}{\partial x_i}\left(K_{ij} \frac{\partial \phi}{\partial x_j}\right) \tag{10.10}$$

P and ϕ are referred to as the state variables of the equations since their distribution characterizes both the flow and the storage components of the aquifer. Obviously, this assumes that the coefficients of the equations (i.e. the properties of the aquifer) are known *a priori*.

Equations (10.9) and (10.10) are the generally adopted equations describing groundwater flow in three dimensions at any point in an aquifer ignoring any sources or sinks. Source and sink terms may be readily added to the right-hand side of these equations. These additions are more satisfactorily illustrated in the following section.

10.2.2 The Hydraulic Approach to Aquifer Modelling

The lateral extent of a regional aquifer is usually much greater than its vertical extent. This disparity between the vertical and horizontal dimensions of many aquifers has resulted in the adoption of two-dimensional equations for the description of regional groundwater flow. This is known as the hydraulic approach to aquifer modelling. In this case, the vertical flow components within the body of the aquifer are effectively ignored, and flow is assumed to be everywhere horizontal. This is acceptable as long as the head or pressure changes due to vertical flow are small compared to those arising from horizontal flow. This is generally true for aquifers at some distance (greater than twice the thickness of the aquifer) from point sources/sinks. In the neighbourhood of point sources such as wells or partially penetrating streams, vertical flow components can be significant. Nevertheless, it is possible to introduce corrections in these zones during the development of the regional groundwater flow equation through the adoption of suitable approximations. One such approximation is discussed in relation to the establishment of appropriate boundary conditions in Section 10.2.3.

The development of the two-dimensional equation of groundwater flow at a point $p(x,y)$ can be undertaken by one of two routes. The first route is to integrate the three-dimensional equation of groundwater flow (equation (10.10)) over the vertical. However, the mathematical treatment of the integration is made somewhat complicated in the case of a phreatic aquifer where the water table represents a moving upper boundary for the integration. The second route is to employ the control volume approach where the control volume extends over the full thickness of the aquifer. This approach is conceptually easier to follow and will be adopted here for the case of a leaky phreatic aquifer with both point and distributed sources and sinks (Figure 10.3). Since all flows are assumed to be essentially horizontal, the potential form of Darcy's law is used in which flow is related to the hydraulic head in the aquifer. The hydraulic head in a phreatic aquifer is defined by the position of the water table, $h(x,y,t)$ [L].

Following the approach adopted to establish the general groundwater flow equation, but assuming that the density is constant, the net volumetric inflow to the control volume centred on point p depicted in Figure 10.3 is given by

$$\delta V \delta x \, \delta y = \delta t \left\{ \frac{Q_x|_{x-\delta x/2,y} - Q_x|_{x+\delta x/2,y}}{\delta x} + \frac{Q_y|_{x,y-\delta y/2} - Q_y|_{x,y+\delta y/2}}{\delta y} \right. $$

$$\left. + q_v + w - \sum_{k=1}^{NWELLS} \frac{P_k}{A_k} I_k(x,y) \right\} \delta x \, \delta y \qquad (10.11)$$

where V is the volume of groundwater stored per unit area of aquifer, [L]
Q_x and Q_y are the components of the vertically integrated horizontal flux, Q, $[L^2T^{-1}]$
q_v is the flux through the base of the aquifer, $[LT^{-1}]$
w is the net recharge to the aquifer per unit area, $[LT^{-1}]$
P_k is the abstraction rate from a well at location (x_k, y_k), $[L^3T^{-1}]$
A_k is the cross-sectional area of the well bore at (x_k, y_k), $[L^2]$
and $I_k(x,y)$ is an indicator function ($= 1$ when (x, y) is within the well at (x_k, y_k), and zero otherwise).

Dividing through by δt, δx and δy and taking the limit as δt, δx and δy tend to zero yields

$$\frac{\partial V}{\partial t} = -\frac{\partial Q_i}{\partial x_i} + q_v + w - \sum_{k=1}^{NWELLS} \frac{P_k}{A_k} I_k(x,y) \qquad (10.12)$$

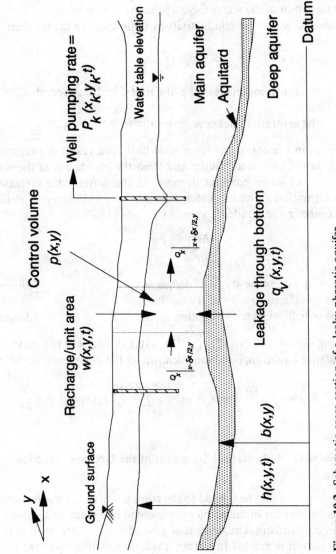

Figure 10.3. Schematic cross-section of a leaky phreatic aquifer

The volumetric flux, $\mathbf{Q}$, can be replaced by its Darcian equivalent:

$$Q_i = - T_{ij} \frac{\partial h}{\partial x_j} \tag{10.13}$$

where T_{ij} is the ijth component of the transmissivity tensor, $\qquad\qquad$ $[L^2 T^{-1}]$
and $\quad h$ is the elevation of the water table above the datum. $\qquad\qquad$ $[L]$

Assuming that the hydraulic conductivity is uniform over depth, then T_{ij} can be defined as follows:

$$T_{ij} = K_{ij}(h - b) \tag{10.14}$$

where K_{ij} $\quad$ is the ijth component of the horizontal hydraulic conductivity
$\qquad\qquad$ tensor, $\qquad\qquad\qquad\qquad\qquad\qquad\qquad\qquad\qquad$ $[LT^{-1}]$
and $\quad (h - b)$ is the saturated thickness of the aquifer. $\qquad\qquad\qquad$ $[L]$

Storage changes in the control volume arise both as a result of pressure changes within the body of the saturated aquifer and from the movement of the water table. The rate of release of water per unit decrease in the water table elevation due to drainage of the formation above the water table is known as the *specific yield*. The rate of change of storage in the aquifer is given by

$$\frac{\partial V}{\partial t} = S \frac{\partial h}{\partial t} \tag{10.15}$$

where $S = S_s(h - b) + S_y$ is the storage coefficient, $\qquad\qquad$ [dimensionless]
$\qquad\quad S_s$ $\quad$ is the specific storativity of the aquifer, $\qquad\qquad$ $[L^{-1}]$
and $\quad S_y$ $\quad$ is the specific yield of the aquifer. $\qquad\qquad\qquad$ [dimensionless]

Thus, combining equations (10.12), (10.13), (10.14) and (10.15) yields the two-dimensional balance equation, commonly known as the *Boussinesq* equation:

$$\frac{\partial}{\partial x_i} \left(K_{ij}(h - b) \frac{\partial h}{\partial x_j} \right) + q_v + w - \sum_{k=1}^{NWELLS} \frac{P_k}{A_k} I_k(x_1, x_2) = S \frac{\partial h}{\partial t} \tag{10.16}$$

10.2.3 The Complete Mathematical Statement of the Groundwater Flow Problem

In this section discussion will be limited to the completion of the mathematical model for groundwater movement of the aquifer represented in Figure 10.3. Whilst equation (10.16) describes the horizontal movement of groundwater at any point in the aquifer, additional information is needed to obtain a solution to this equation at all points over the aquifer domain. First, boundary conditions must be specified to define the link between the aquifer and the surrounding hydrogeological environment. Second, the head distribution over the whole aquifer must be specified at some initial time point. Both the boundary and the initial conditions to be assigned to the model region must be known *a priori*.

Initial Conditions

As noted previously, an initial head distribution over the entire aquifer domain is required at an initial time, usually $t = 0$. This is conveniently described mathematically as:

$$h(x,y,0) = H_0(x,y) \qquad (10.17)$$

where $H_0(x,y)$ is the initial head distribution $\qquad\qquad$ [L]

Lateral Boundary Conditions

The simplest boundary conditions that can be assigned are either (i) fixed head or (ii) fixed flux.

Fixed heads are denoted on the line of a boundary, B, by

$$h(x,y,t) = H_B(x,y) \qquad \text{on} \quad B \qquad (10.18)$$

where H_B is the fixed head distribution on the boundary.

Fixed-head boundary conditions are frequently used to represent connections between the aquifer and rivers, lakes or spring lines.

Fixed fluxes are usually defined normal to the line of the boundary and are thus denoted on the line of the boundary, B, by

$$Q_n = Q_B(x,y) \qquad \text{on} \quad B \qquad (10.19)$$

where Q_n is the flux normal to the boundary per unit width of aquifer. $\qquad$ [LT^{-1}]

The commonest flux boundary condition is given when $Q_n = 0$. This is a no-flow boundary representing the presence of an impermeable boundary or simply a line across which no flow takes place such as a streamline.

In the case of a fixed head boundary, it is assumed that the water body corresponding to the fixed head fully penetrates the aquifer. Where rivers and streams only partially penetrate the aquifer and where ignoring the vertical flow components in the aquifer can lead to significant errors in the predicted head distributions, an effective resistance can be introduced to account for the additional head losses adjacent to the stream. This is illustrated diagrammatically in Figure 10.4.

More complex boundary equations can be obtained by extension of these basic conditions to allow for time-varying distributions, or interaction between the boundary conditions and the groundwater conditions in the aquifer. However, equations describing such conditions are generally specific to the particular problem being solved and, thus, will not be discussed here.

Once the general balance equation, the boundary conditions and the initial conditions have been defined, the only remaining requirements are to define the distributions of each of the model parameters (i.e. the storage coefficients, hydraulic conductivity and aquifer boundaries) over the aquifer domain for the time period of interest. Having completed this exercise, the mathematical model is completely defined.

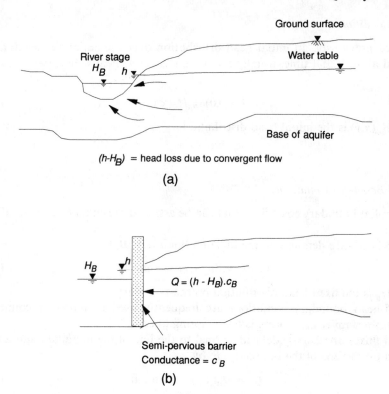

$(h-H_B)$ = head loss due to convergent flow

(a)

$$Q = (h - H_B).c_B$$

Semi-pervious barrier
Conductance = c_B

(b)

Figure 10.4. Semi-pervious river boundary condition: (a) cross-section of river–aquifer interface; (b) conceptual approximation of boundary condition

It is important to note that the flux distributions obtained over the domain through the equations defined in this section are smoothed compared to the actual flux distributions observed in the aquifer. The effect of this smoothing on the development of the transport equations is discussed in more detail in Section 10.3.

10.3 CONTAMINANT TRANSPORT

Substances of widely differing natures are transported in groundwater. The processes affecting the transport of these substances are very complex, and the degree of complexity tends to increase with the number of chemical species present. In addition to physical transport processes, biological and chemical interactions can occur, both with the rock matrix and with other species. In order to identify and study the physical mechanisms involved in groundwater contaminant transport, attention will be restricted initially to the simplest case, namely that of a single species which, in the context of groundwater transport, is chemically and biologically inert and which occurs in concentrations which do not substantially affect the density or viscosity of the groundwater. Such a substance is often referred to as an ideal or conservative tracer. Throughout this chapter, the *concentration c* $[ML^{-3}]$ of a chemical species in an aquifer is taken to be defined as the mass of the species per unit volume of groundwater.

10.3.1 Transport by Advection

In determining the fate of a contaminant in groundwater, the rate of migration of the water is of fundamental importance. Since flow models calculate average groundwater fluxes, it is useful to introduce the concept of an average water velocity. Not all the water in an REV travels at the same speed, indeed a certain amount of water is attached by strong forces of molecular attraction to the surfaces of the rock matrix and is effectively immobile. Similarly, some water may be contained in small pores and fissures which greatly restrict its motion. The *advective velocity*, $\mathbf{u}$ $[LT^{-1}]$ of the water at a point is defined to be the average velocity of the movable groundwater in an REV centred at that point. Central to the evaluation of the velocity is the concept of effective porosity. The *total* or *bulk porosity* [dimensionless] of a volume of aquifer is given by

$$n = \frac{\text{total volume of void space}}{\text{total volume}} \tag{10.20}$$

The *kinematic* or *effective* porosity [dimensionless] of a volume of aquifer is defined by

$$n_e = \frac{\text{volume occupied by movable water}}{\text{total volume}} \tag{10.21}$$

If ξ [dimensionless] is defined as the fraction of the porewater that is movable, the relationship between the total and effective porosities can be expressed as $n_e = \xi n$. Both notations will be used in the following sections. The effect of immobile water on mass transport will be more fully discussed in Section 10.3.5.

The specific discharge, $\mathbf{q}$ $[LT^{-1}]$ is the groundwater volume flux, i.e. the volume of water crossing a unit area of aquifer in a unit time. Only a fraction of a cross-section of aquifer is open to water flow, and it is assumed that the effective porosity provides a direct measure of that fraction. It follows that the relationship between specific discharge and advective velocity is given by

$$\mathbf{u} = \frac{1}{n_e} \mathbf{q} \tag{10.22}$$

Some authors use the term 'Darcy velocity' when referring to specific discharge. This is potentially confusing and is best avoided.

Transport solely by advection entails that tracers migrate with the same velocity as the water and that the fate of pollutants can therefore be determined from the patterns of fluid flow through the aquifer. With the exception of Section 10.3.5, it will be assumed that all the water in the aquifer is mobile and hence that the total and effective porosities are the same. Given that the contaminant concentration is c and the effective porosity is n_e, the mass of contaminant, per unit volume of aquifer, that can be transported by advective processes is $n_e c$ and so the advective mass flux, $\mathbf{J}^a$, expressed as a mass per unit area of aquifer per unit time $[ML^{-2}T^{-1}]$ is given by

$$\mathbf{J}^a = n_e c \mathbf{u} \qquad (10.23)$$

10.3.2 Transport by Dispersion

Measured contaminant breakthrough curves exhibit spreading of the contaminant in the direction of flow. Lateral spreading at right angles to the direction of flow is also observed. Furthermore, in the absence of advective transport, contaminant migration still occurs. This spreading of the contaminant away from the position determined by the advective velocity is called *hydrodynamic dispersion* and is governed by physical processes that occur at a variety of length scales.

At the microscopic scale the mixing of solutions of different concentrations takes place by *molecular diffusion* which is caused by the Brownian motion of water molecules. This process is isotropic and is governed by Fick's law which, for diffusion in free water, can be written

$$J_i^d = - d_0 \frac{\partial c}{\partial x_i} \qquad (10.24)$$

where J_i^d is the ith component of the mass flux due to diffusion, $\quad [ML^{-2}T^{-1}]$
and $\quad d_0$ is the free water diffusion coefficient. $\qquad\qquad [L^2T^{-1}]$

In porous media, the diffusion coefficient must also account for the geometry of the pores. This is achieved by introducing a tortuosity factor, χ [dimensionless]. Furthermore, in equation (10.24), the diffusive flux is expressed as a mass per unit area of fluid per unit time, but for use in the groundwater mass balance equation, the flux needs to be expressed as a mass per unit area of aquifer per unit time. This is achieved by multiplying d_0 by the total porosity. These two adjustments allow the definition of an *effective* or *intrinsic coefficient of molecular diffusion, d_e,* by

$$d_e = d_0 n \chi \qquad (10.25)$$

and so, in the context of diffusive transport in groundwater, Fick's law can be written

$$J_i^d = - d_e \frac{\partial c}{\partial x_i} \qquad (10.26)$$

Above the micro-scale, but within the REV, the spreading of a contaminant is brought about by advective processes. This spreading is referred to as *mechanical dispersion* and is due to spatial variations in water velocity which depend on the structure of the aquifer material and the advective velocity of the water. Within individual pores, velocity variations arise due to boundary layer effects and the variability of pore geometry. At the pore scale, variations are caused by differences in average pore diameter and by water being forced around individual grains of aquifer material or small areas of low permeability. Mechanical dispersion is represented by a Fickian model:

$$J_i^m = - D_{ij}^m \frac{\partial c}{\partial x_j} \tag{10.27}$$

where J_i^m is the ith component of the mass flux due to mechanical
dispersion expressed as a mass per unit area of aquifer per
unit time, $[ML^{-2}T^{-1}]$
and D_{ij}^m is the ijth component of the mechanical dispersion tensor. $[L^2T^{-1}]$

It should be noted here that mechanical dispersion is an anisotropic process that is dependent not only on the nature of the porous medium, but also the mean direction of flow. The mechanical dispersion tensor is a function of the water velocity vector and two constants α_L and α_T, the *longitudinal* and *transverse dispersivities* [L] respectively. The magnitudes of these dispersivities reflect the amount of spreading in the direction of flow and at right angles to it (α_L is typically an order of magnitude greater than α_T). For an isotropic porous medium, the components of the mechanical dispersion tensor can be shown to be

$$D_{ij}^m = n_e \left\{ \alpha_T u \delta_{ij} + (\alpha_L - \alpha_T) \frac{u_i u_j}{u} \right\} \tag{10.28}$$

where u is the magnitude of the average advective velocity vector $\mathbf{u}$ $[LT^{-1}]$
and δ_{ij} is the Kronecker delta ($\delta_{ij} = 1$ if $i = j$, and 0 otherwise).

If under uniform flow conditions the x_1 coordinate axis is aligned with the direction of flow, the mechanical dispersion tensor simplifies to

$$n_e u \begin{bmatrix} \alpha_L & 0 & 0 \\ 0 & \alpha_T & 0 \\ 0 & 0 & \alpha_T \end{bmatrix} \tag{10.29}$$

The total hydrodynamic dispersive flux, $\mathbf{J}^D$, is the sum of the mechanical dispersive flux and the molecular diffusive flux. Since both these component fluxes are assumed to be Fickian in nature, the hydrodynamic flux is also assumed to obey a Fickian law

$$J_i^D = - D_{ij} \frac{\partial c}{\partial x_j} \tag{10.30}$$

where J_i^D is the ith component of the hydrodynamic dispersive flux, $[ML^{-2}T^{-1}]$
and $D_{ij} = D_{ij}^m + d_e$ is the ijth component of the hydrodynamic dispersion
coefficient, $\mathbf{D}$. $[L^2T^{-1}]$

Having established the form of the total dispersive flux, it is now possible to develop the equation describing transport by advection and dispersion in porous media. Here, it is assumed again that all water in the aquifer is mobile and hence that the effective porosity and total porosity are the same. It was seen in Section 10.2.1 that for an incompressible fluid the rate of change of mass per unit volume of medium is equal to minus the divergence of the mass flux vector. Given that the contaminant concentration is c and the total porosity n, the total mass of contaminant per unit volume of aquifer is nc, and so by considering the total mass flux due to advection and dispersion the contaminant mass balance equation is

$$\frac{\partial}{\partial t}(nc) = \frac{\partial}{\partial x_i}\left(D_{ij}\frac{\partial c}{\partial x_j} - n_e c u_i\right) \tag{10.31}$$

Equation (10.31) is known as the *advection–dispersion equation*. Some of the literature on this subject defines the dispersion coefficient using equations in which the dispersive flux is defined as a mass per unit area of fluid per unit time. This leads to a dispersion coefficient, $\mathbf{D}'$, given by

$$\mathbf{D}' = \frac{1}{n_e}\mathbf{D} \tag{10.32}$$

The reader is, therefore, advised to be clear about which dispersion coefficient is being used in a particular application.

10.3.3 Production and Decay of the Transported Species

It is often the case that chemical compounds transported in groundwater do not act like conservative tracers, but decompose due to chemical or microbiological reactions into other species, or are formed as a product of similar reactions. Identifying the nature of these reactions and quantifying the speed at which they take place is the subject of significant research. However, at the aquifer scale, these reactions are usually modelled by first-order kinetics, that is, the reaction rate is assumed to be a linear function of the concentration of the species. For example, the decay of a chemical species can often be regarded as obeying the law

$$\frac{\partial c}{\partial t} = -\lambda c \tag{10.33}$$

where λ is the reaction rate constant. $\qquad [T^{-1}]$

This will also be recognized as the law governing the decay of radioactive substances. Given that the mass of contaminant per unit volume of aquifer is nc, it follows that λnc must appear as a sink term in the total contaminant mass balance equation. This gives

$$\frac{\partial}{\partial t}(nc) = \frac{\partial}{\partial x_i}\left(D_{ij}\frac{\partial c}{\partial x_j} - n_e c u_i\right) - \lambda nc \tag{10.34}$$

When species decay they produce 'daughter' products which may themselves decay further. Thus a chain of products is formed, and the transport of each species may be modelled in parallel. In such cases mass balance equations for each species in the chain must be developed in which the sink term representing decay in the equation of the 'parent' species becomes a source term in the equation of the 'daughter' product. This aspect of transport modelling will not be pursued further here.

10.3.4 Adsorption

So far, discussion has centred on transported substances which do not react with the rock matrix. However, many substances do react in this way. In this section the process of adsorption onto the surface of the aquifer material is considered. Adsorptive and desorptive processes which occur at the rock surface can have a significant effect on the timing of contaminant breakthrough. In constructing the total mass balance equation for such a contaminant it is useful to divide conceptually the total mass of contaminant into the mass in solution and the mass adsorbed on the rock surface. The mass of contaminant in solution, expressed as a mass per unit volume of aquifer, is nc. If F [dimensionless] is the mass of contaminant adsorbed per unit mass of solid aquifer material, then the total mass of adsorbed contaminant per unit volume of aquifer is $(1 - n)\rho_s F$, where ρ_s is the density of the rock matrix. Thus, remembering that the advective and dispersive mass fluxes are functions only of the mass in solution, the total contaminant mass balance equation including decay is

$$\frac{\partial}{\partial t}\{nc + (1 - n)\rho_s F\} = \frac{\partial}{\partial x_i}\left(D_{ij}\frac{\partial c}{\partial x_j} - n_e c u_i\right) - \lambda\{nc + (1 - n)\rho_s F\} \quad (10.35)$$

For simplicity, it has been assumed in equation (10.35) that the rate of contaminant decay is the same in the adsorbed and non-adsorbed phases, as in the case of radioactive decay. However, the basic ideas presented here are easily modified to include differing decay rates. Equation (10.35) has two state variables, F and c, and so to solve the equation it is necessary to obtain a relationship between them. Such a relationship is called an *adsorption isotherm* which expresses the mass balance of the pollutant on the rock surface (the term isotherm derives from the fact that the relationship is valid only at a constant temperature). The form of the isotherm depends on the nature of the transported substance and of the aquifer material, and on the conceptual model of the adsorptive processes. *Equilibrium isotherms* are based on the assumption that an equilibrium exists at all times between the concentration of the contaminant in the pore water and the concentration of the contaminant adsorbed on the rock surface. In other words, it is assumed that the time taken for the adsorbed concentration to change in response to a change in the pore water concentration is small. *Non-equilibrium isotherms*, on the other hand, are based on the assumption that adsorption and desorption take place over long enough periods of time to warrant modelling as a function of time. The simplest relationship between F and c is the linear equilibrium isotherm:

$$F = K_d c \quad (10.36)$$

where K_d is known as the distribution coefficient. $[M^{-1}L^3]$

This is often assumed to be valid when the solute concentration is low. In this case, equation (10.35) can be written

$$\frac{\partial}{\partial t}(Rnc) = \frac{\partial}{\partial x_i}\left(D_{ij}\frac{\partial c}{\partial x_j} - n_e c u_i\right) - \lambda Rnc \quad (10.37)$$

where

$$R = 1 + \left(\frac{1 - n}{n}\right)\rho_s K_d \tag{10.38}$$

is known as the retardation factor [dimensionless]. The effect of R on contaminant breakthrough can be seen by considering the case in which R is constant. Dividing through equation (10.37) by R produces an equation in the form of the advection–dispersion equation with an apparent velocity of $\mathbf{u}/R$. Since R is greater than 1, this apparent velocity is less than the advective velocity and so the arrival of the adsorbing contaminant at a sampling point will be later than that of a non-adsorptive tracer. However, if the advective velocity is uniform, the shape of the breakthrough curve at the point will be the same for both kinds of contaminant.

A number of other isotherms are commonly used; in particular the nonlinear Langmuir isotherm

$$F = \frac{K_1 c}{1 + K_2 c} \qquad K_1, K_2 > 0 \tag{10.39}$$

and the Freundlich isotherm

$$F = K_1 c^{1/m} \qquad K_1 > 0, m \geq 1 \tag{10.40}$$

The simplest non-equilibrium isotherm, also attributed to Langmuir, is

$$\frac{\partial F}{\partial t} = K_1 c. \tag{10.41}$$

In cases in which the adsorption isotherm does not express F explicitly as a function of c, it is necessary to solve the total mass balance equation and the isotherm as a system of two separate equations.

10.3.5 The Dual Porosity Model

Contaminant breakthrough curves often display longer tails than are produced by the models that have been described above. This has been explained using the conceptual model of a dual porosity medium. In this model the water in the aquifer is thought of as being divided into mobile and immobile fractions. As contaminated water migrates through a clean aquifer, some of the contaminant will diffuse into regions containing effectively immobile water. If the contaminant source is then removed, clean water passing through the aquifer will become polluted by the contaminant diffusing out of the immobile region. Applying this model and ignoring adsorption and decay, the pollutant mass balance equation for the mobile and immobile regions is

$$\frac{\partial}{\partial t}\{\xi n c + (1 - \xi)n c^*\} = \frac{\partial}{\partial x_i}\left(D_{ij}\frac{\partial c}{\partial x_j} - \xi n c u_i\right) \tag{10.42}$$

where c^* is the contaminant concentration in the immobile water. [ML^{-3}]

The exchange of mass between the mobile and immobile regions is regarded as a diffusive process. A first-order approximation to this process is often used to define the relationship between the state variables c and c^* which must be established in order to be able to solve equation (10.42) for c. This produces a mass balance equation for the immobile region of the form

$$\frac{\partial}{\partial t}\{(1 - \xi)nc^*\} = \alpha(c - c^*) \tag{10.43}$$

where α is the diffusive exchange coefficient. $[T^{-1}]$

The dual porosity model can be extended to include the processes of adsorption and decay by conceptually dividing the adsorption sites on the rock surface into those sites available to the contaminant in the mobile water and those sites surrounded by immobile water. If f [dimensionless] is defined to be the fraction of the total number of adsorption sites that are available to the contaminant in the mobile water, then the total mass of contaminant per unit volume of aquifer, M, is given by

$$M = \xi nc + (1 - \xi)nc^* + f\rho_s(1 - n)F + (1 - f)\rho_s(1 - n)F^* \tag{10.44}$$

Here, F and F^* are the masses of pollutant per unit mass of solid aquifer material in the sites adjacent to the mobile and immobile water respectively. Therefore, the total mass balance equation, including a decay term, can be written

$$\frac{\partial M}{\partial t} = \frac{\partial}{\partial x_i}\left(D_{ij}\frac{\partial c}{\partial x_j} - \xi ncu_i\right) - \lambda M \tag{10.45}$$

Assuming that the same linear adsorption isotherm can be applied in both regions, this can be rearranged to give

$$\frac{\partial}{\partial t}\{R_1\xi nc\} + \frac{\partial}{\partial t}\{R_2(1 - \xi)nc^*\}$$

$$= \frac{\partial}{\partial x_i}\left(D_{ij}\frac{\partial c}{\partial x_j} - \xi ncu_i\right) - \lambda\{R_1\xi nc + R_2(1 - \xi)nc^*\} \tag{10.46}$$

where

$$R_1 = 1 + \frac{(1 - n)}{\xi n}f\rho_s K_d \quad \text{and} \quad R_2 = 1 + \frac{(1 - n)}{(1 - \xi)n}(1 - f)\rho_s K_d \tag{10.47}$$

Similarly, the mass balance equation for the immobile region is

$$\frac{\partial}{\partial t}\{R_2(1 - \xi)nc^*\} = \alpha(c - c^*) - \lambda R_2(1 - \xi)nc^* \tag{10.48}$$

10.3.6 Initial and Boundary Conditions

In Section 10.2.3 it was seen that the complete mathematical description of the groundwater flow problem requires the specification of a set of initial and boundary conditions. The same is true in the case of contaminant transport problems, and the types of conditions in both classes of problem are essentially the same. First, the initial contaminant concentration distribution throughout the aquifer must be specified.

Second, boundary conditions, which may be time varying, must be imposed on the entire boundary of the modelled region. These can be of one of two kinds. Prescribed concentration boundary conditions assign the value of the contaminant concentration on a section of the boundary at any given time. Prescribed flux boundary conditions give the value of the total contaminant mass flux normal to the boundary of the modelled region.

10.4 NUMERICAL METHODS

Analytical solutions to groundwater flow and transport problems are available only for a handful of simple cases. These solutions can be useful in some circumstances as first approximations to real problems and are invaluable in testing numerical methods. The fact remains that almost all real pollution problems are too complex for analytical treatment and so numerical methods of solution are required. As mentioned in Section 10.3, the transport problems considered are ones in which neither the density nor the viscosity of the pore water is affected by the solute concentration. This greatly simplifies the solution of transport problems, but it must be remembered that in modelling the transport of some contaminants, density effects cannot be ignored and techniques beyond the scope of this chapter must be employed. If the effects of the changes in density and viscosity of the porewater are insignificant, then the flow and transport equations can be solved separately. Solving the flow problem produces a field of head values which can be translated into specific discharge or groundwater velocity fields which are then used as input to the transport problem.

In this section, numerical methods for solving groundwater flow and transport problems are outlined and some of the inherent difficulties are discussed. For simplicity, the methods are described here in terms of finite differences, but much work has been carried out on the development of finite element solutions, and a detailed introduction is provided by Huyakorn and Pinder (1983). A number of finite difference approximations to the relevant partial differential equations can be constructed, but not all such approximations produce good solution techniques. Any useful solution procedure must have the properties of convergence and stability. A method is *convergent* if the numerical results approach the analytical solution as the spatial and temporal increments, Δx and Δt, tend to zero. A method is said to be *stable* if an error at one stage in the calculation is not propagated through subsequent stages but is eventually damped out. In cases where considerable computational expense is incurred in solving a problem, methods which converge quickly are particularly valuable.

10.4.1 Steady-state Flow Calculations

The steady-state groundwater flow equation for a confined aquifer with no sources or sinks is

$$\frac{\partial}{\partial x_i}\left(T_{ij}\frac{\partial \phi}{\partial x_j}\right) = 0 \tag{10.49}$$

Since the transmissivity tensor is symmetric, the principal directions of anisotropy are orthogonal and, aligning the coordinate axes with these directions, the two-dimensional form of the equation can be written

$$\frac{\partial}{\partial x}\left(T_x \frac{\partial \phi}{\partial x}\right) + \frac{\partial}{\partial y}\left(T_y \frac{\partial \phi}{\partial y}\right) = 0 \tag{10.50}$$

In equation (10.50), the non-zero elements, T_{11} and T_{22}, of the transmissivity tensor have been replaced by T_x and T_y respectively. The finite difference approximation to this equation can be developed using the grid in Figure 10.5. First, approximations to the outer derivatives are made giving

$$\frac{\left(T_x \frac{\partial \phi}{\partial x}\right)_{i+1/2,j} - \left(T_x \frac{\partial \phi}{\partial x}\right)_{i-1/2,j}}{\Delta x} + \frac{\left(T_y \frac{\partial \phi}{\partial y}\right)_{i,j+1/2} - \left(T_y \frac{\partial \phi}{\partial y}\right)_{i,j-1/2}}{\Delta y} = 0 \tag{10.51}$$

The bracketed terms in equation (10.51) are then approximated to give the full finite difference representation:

$$(T_x)_{i+1/2,j}\frac{(\phi_{i+1,j} - \phi_{i,j})}{(\Delta x)^2} - (T_x)_{i-1/2,j}\frac{(\phi_{i,j} - \phi_{i-1,j})}{(\Delta x)^2}$$

$$+ (T_y)_{i,j+1/2}\frac{(\phi_{i,j+1} - \phi_{i,j})}{(\Delta y)^2} - (T_y)_{i,j-1/2}\frac{(\phi_{i,j} - \phi_{i,j-1})}{(\Delta y)^2} = 0 \tag{10.52}$$

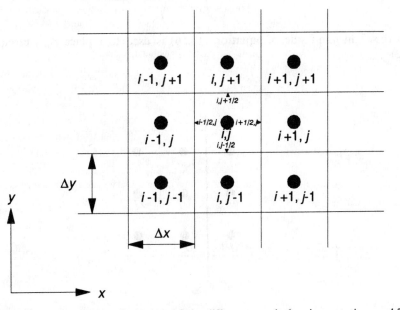

Figure 10.5. Segment of a two-dimensional finite difference mesh showing notation used for the nodal equations

The transmissivity terms such as $(T_x)_{i+1/2,j}$ represent average values of, in this case, $(T_x)_{i+1,j}$ and $(T_x)_{i,j}$. The manner in which this average is calculated is a matter of present research interest, but it is common practice to use the harmonic mean

$$(T_x)_{i+1/2,j} = \frac{2(T_x)_{i+1,j}(T_x)_{i,j}}{(T_x)_{i+1,j} + (T_x)_{i,j}} \tag{10.53}$$

Fixed-head boundary conditions are imposed by setting the head values at the appropriate nodes of the grid. Equation (10.52) defined for all nodes in the domain, together with the boundary values, define a set of linear equations which can be solved numerically to give head values at each node.

The process by which fixed flux boundary conditions are imposed is more complex. Figure 10.6 shows the edge of a finite difference grid and it is supposed that a boundary condition of

$$T_x \frac{\partial \phi}{\partial x} = f \quad \text{(constant)} \tag{10.54}$$

is required. In order to use a central difference estimate of the derivative, an extra node $(0, j)$ is employed which is outside the modelled region. This gives

$$\left(\frac{\partial \phi}{\partial x} \right)_{1,j} \approx \frac{\phi_{1,j} - \phi_{0,j}}{\sigma \Delta x} \tag{10.55}$$

that is

$$\phi_{0,j} \approx \phi_{1,j} - \frac{2f\Delta x}{T_x} \tag{10.56}$$

Thus, the right-hand side of equation (10.56) is used to replace $\phi_{0,j}$ in equation (10.52).

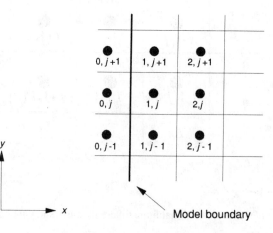

y

x

Model boundary

0, j+1   1, j+1   2, j+1

0, j   1, j   2, j

0, j-1   1, j-1   2, j-1

Figure 10.6. Segment of a two-dimensional finite difference mesh showing boundary nodes

Flow in an unconfined aquifer is governed by equation (10.16). In the steady-state case, with no source or sink terms, this can be written

$$\frac{\partial}{\partial x}\left(K_x(h-b)\frac{\partial h}{\partial x}\right) + \frac{\partial}{\partial y}\left(K_y(h-b)\frac{\partial h}{\partial y}\right) = 0 \qquad (10.57)$$

As in the case of the confined aquifer solution, a finite difference approximation to equation (10.57) can be constructed without difficulty, but any such approximation will contain terms which are not linear functions of h. Consequently, the system of equations defined by the finite difference approximation and the boundary conditions are non-linear. This entails the use of more sophisticated solution techniques. There are a number of ways of dealing with this complication. A commonly used technique is to linearize equation (10.57) and to employ a simple iterative solution procedure. At the start of each time step, an initial estimate of the terms $K(h-b)$ is made at each node. For the first iteration, these values are regarded as constants. The resulting finite difference equations are linear and these are solved to give values of h at each node. New estimates of $K(h-b)$ are then made using the calculated values of h, and the solution process is repeated. At each step in the iterative process, a measure of the change in the solution from the previous time step is calculated, and when this is acceptably small the process is terminated.

10.4.2 Transient Flow and Diffusion Calculations

The transient flow equation for a confined aquifer with no sources or sinks is

$$\frac{\partial}{\partial x_i}\left(T_{ij}\frac{\partial \phi}{\partial x_j}\right) = S\frac{\partial \phi}{\partial t} \qquad (10.58)$$

and the equation describing transport by diffusion only in an incompressible aquifer is

$$\frac{\partial}{\partial x_i}\left(D_{ij}\frac{\partial c}{\partial x_j}\right) = n\frac{\partial c}{\partial t} \qquad (10.59)$$

These are mathematically similar, and in this section the methods of solution of these equations are shown in the context of the transport problem. It should be clear, however, that the results apply equally to the flow problem. To understand the techniques involved in solving the diffusive transport problem numerically, it is useful to consider cases which are readily amenable to investigation. Thus, in this section and Section 10.4.3, attention is restricted to the one-dimensional case, with a constant dispersion coefficient. In this case the governing equation is

$$\frac{\partial c}{\partial t} = \frac{D}{n}\frac{\partial^2 c}{\partial x^2} \qquad (10.60)$$

A number of finite difference discretizations of this equation are possible. First, an explicit formulation is given by

$$\frac{c_i^{m+1} - c_i^m}{\Delta t} = \frac{D}{n}\frac{(C_{i+1}^m - 2c_i^m + c_{i-1}^m)}{(\Delta x)^2} \qquad (10.61)$$

where subscripts and superscripts refer to spatial and time increments respectively.

This can be written

$$c_i^{m+1} = c_i^m + \frac{D\Delta t}{n(\Delta x)^2}(c_{i+1}^m - 2c_i^m + c_{i-1}^m) \tag{10.62}$$

This formulation has the advantage that the concentration at any location in space can be calculated directly from the known concentrations at the previous time step. It can be shown that this method is both convergent and stable provided that

$$\frac{D\Delta t}{n(\Delta x)^2} \le \frac{1}{2} \tag{10.63}$$

The mesh size Δx is usually determined by physical considerations or by computational limitations. The condition is, therefore, usually interpreted as a restriction on the size of the time step, namely

$$\Delta t \le \frac{n(\Delta x)^2}{2D} \tag{10.64}$$

This method is simple and requires little computational effort at each time step, but in some applications the restriction on the size of the time step can render the method impractical. In cases in which there are temporal discontinuities in the boundary conditions, such as at the start of the injection of a slug of contaminant, oscillations of the numerical result about the analytical solution are often observed. In order to minimize these oscillations, a fine finite difference mesh is required, but this entails the use of small time steps. The effect of reducing the mesh size can be severe; if Δx is halved the time step must be reduced by a factor of four, and thus eight times the number of calculations are required in total. The problem of computational expense is even more critical in two- and three-dimensional problems. Not only are there more calculations necessary due to the increased number of mesh elements, but the stability and convergence criteria are even more restrictive.

In an attempt to overcome the problems of stability and convergence, the *fully implicit* method is often used. This is given by

$$\frac{c_i^{m+1} - c_i^m}{\Delta t} = \frac{D}{n}\frac{(c_{i+1}^{m+1} - 2c_i^{m+1} + c_{i-1}^{m+1})}{(\Delta x)^2} \tag{10.65}$$

The method can be shown to be convergent and unconditionally stable, thus allowing the use of larger time steps. However, it is not possible to calculate the concentration at a point explicitly from point concentrations at the previous time step, and so all point concentrations at a given time step must be calculated simultaneously. This entails the solution of a system of linear equations at each time step. Fortunately, the system of equations produced by this method can be solved rapidly and accurately. Nevertheless, there is a decision to be made between using many small time steps at which simple calculations are performed as in the explicit method, and using larger time steps entailing more complex calculations at each step.

Both the methods discussed so far employ one-sided difference approximations to the time derivative and central difference approximations to the spatial derivative.

These approximations have errors of the order Δt and $(\Delta x)^2$ respectively. The use of this approximation to the time derivative can involve an associated error of the same order of magnitude as the dispersive term in the original equation. This gives rise to a global error in the numerical solution which shows up as an enhanced dispersion. For this reason this error is generally referred to as *numerical dispersion*. The inconsistency in the order of the error terms can be overcome by using central difference approximations throughout. This can be achieved by evaluating the time and spatial derivatives at the time step $m + \frac{1}{2}$ (that is half-way between the beginning of time step m and the beginning of time step $m + 1$). The time derivative is given by a central difference approximation with increment $\Delta t/2$. This has an associated error of order $(\Delta t/2)^2$. The spatial derivative is calculated by taking the arithmetic mean of the central difference approximations at the beginning of time step m and the beginning of time step $m + 1$. This is the Crank–Nicholson method and is given by

$$\frac{c_i^{m+1} - c_i^m}{\Delta t} = \frac{D}{2n(\Delta x)^2} (c_{i+1}^{m+1} - 2c_i^{m+1} + c_{i-1}^{m+1} + c_{i+1}^m - 2c_i^m + c_{i-1}^m) \quad (10.66)$$

This method can be shown to be unconditionally stable for this problem and to converge more quickly than the fully implicit method. It also has the property that numerical dispersion is kept to a minimum.

10.4.3 Stability of Central Difference Approximation Techniques for Transport Calculations

In diffusive transport problems, the Crank–Nicholson scheme has the desirable properties of rapid convergence and stability, but difficulties can arise in some problems involving advection. In the numerical treatment of transport problems, the relative significance of advection and dispersion is measured by the dimensionless *mesh Péclet number* (Pe_Δ) given by

$$Pe_\Delta = \frac{q\Delta x}{D} = \frac{u\Delta x}{D'} \quad (10.67)$$

Thus, high mesh Péclet numbers indicate advection dominated transport. In finite difference solutions to the problem governed by the equation

$$\frac{\partial c}{\partial t} = \frac{D}{n_e} \frac{\partial^2 c}{\partial x^2} - u \frac{\partial c}{\partial x} \quad (10.68)$$

the Crank–Nicholson scheme remains stable provided the mesh Péclet number is less than 2. In transport problems involving higher mesh Péclet numbers, oscillations can occur in the solution as shown by the breakthrough curve in Figure 10.7. Thus, in highly advective situations there is a choice to be made between the Crank–Nicholson scheme, which is unstable, and the fully implicit scheme which exhibits numerical dispersion (i.e. which is inaccurate). The mesh Péclet number can be reduced by defining the finite difference mesh, but the computational expense incurred in such a refinement can be prohibitive, especially in two- or three-dimensional problems.

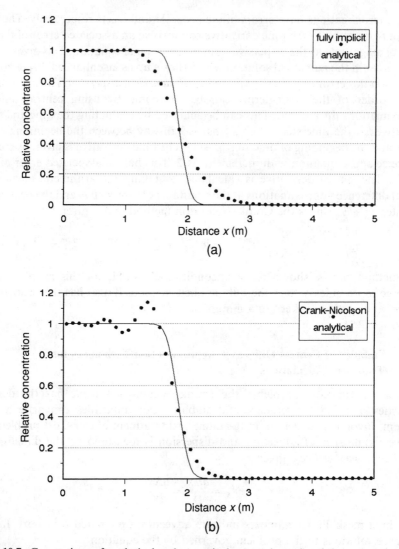

Figure 10.7. Comparison of analytical and numerical contaminant breakthrough curves for a constant input concentration: (a) fully implicit solution for $Pe = 18.5$, $t = 8$ days; (b) Crank-Nicholson solution for $Pe = 18.5$, $t = 8$ days

Finite element solutions of the transport problem are also subject to mesh Péclet number instability, but they tend to be able to cope with higher mesh Péclet numbers than 2 before serious instabilities arise. The difficulties encountered in solving advection dominated transport problems by central finite difference techniques have led to the development of alternative solution strategies. Two of these are outlined in the following sections.

10.4.4 The Transport Problem Using the Random Walk Method

The two-dimensional 'random walk' model developed by Prickett *et al.* (1981) is based on the form of the analytic solution of the one-dimensional advective-dispersive transport problem with an instantaneous slug injection of contaminant of concentration c_0. For an incompressible aquifer, steady flow and a constant dispersion coefficient, the governing equation of this problem is

$$\frac{\partial c}{\partial t} + u \frac{\partial c}{\partial x} = D' \frac{\partial^2 c}{\partial x^2} \tag{10.69}$$

where $D' = D/n$.

Assuming that $D' = \alpha'_L u$, the analytical solution is given by

$$\frac{c}{c_0} = \frac{1}{\sqrt{(4\pi\alpha'_L t)}} \exp\left[-\frac{(x - ut)^2}{4\alpha'_L t} \right] \tag{10.70}$$

This has the same form as the normal distribution function $Z(x)$ with mean μ and standard deviation σ:

$$Z(x) = \frac{1}{\sqrt{(2\pi\sigma^2)}} \exp\left[-\frac{(x - \mu)^2}{2\sigma^2} \right] \tag{10.71}$$

It follows that the contaminant distribution at any time t has the same shape as the normal distribution with mean ut and variance $2\alpha_L ut$.

The model employs a fixed rectangular grid superimposed on the modelled domain. A large number of 'particles' are introduced into this domain and each particle is assigned a mass of contaminant equal to the total contaminant mass in the domain divided by the number of particles. The number of particles in a particular grid element is determined by the initial contaminant distribution. Advective transport is represented by moving the particles at each time step with the advective velocity which has been determined from flow calculations. Dispersion is represented by a random movement of each particle at the end of the time step. For longitudinal dispersion this is achieved by moving each particle, in the direction of the advective velocity, a distance equal to a number sampled from the normal distribution with mean zero and variance $2\alpha_L u\Delta t$. This distribution is chosen since adding m samples from a normal distribution with mean zero and variance $2\alpha_L u\Delta t$ produces a normal distribution with mean zero and variance $2\alpha_L um\Delta t$ as required. Lateral dispersion is treated in a similar manner. A limit on the magnitude of the sampled number is imposed to minimize the possibility of producing very large, non-physical displacements of contaminant. After the required number of time steps, the mass of contaminant in each grid element is calculated simply by summing the masses on the individual particles in that element. The distribution of masses can be transformed into an approximate concentration distribution if required.

The advantages of this type of model include the elimination of numerical dispersion, unconditional stability and conceptual simplicity. The model also possesses the unique advantage that the solutions are additive in the sense that the results from two distinct runs of the program can be combined to produce a more accurate approximation of the concentration field. However, the accuracy of the method is strongly dependent upon the density of particles, and the number of particles required to ensure a high density throughout the modelled region can be very large, especially in the case of divergent flows. Consequently, considerable computing power is required in order to employ this method effectively.

10.4.5 The Transport Problem Using Eulerian–Lagrangian Methods

A number of the solution techniques developed to overcome the inadequacies of central difference methods are based on Lagrangian or mixed Eulerian–Lagrangian principles. Eulerian methods for solving groundwater transport problems such as those described in 10.4.2 are based on the calculation of the solute concentration at fixed points in space. Lagrangian methods, on the other hand are based on the calculation of the mass of solute associated with a hypothetical fluid particle as it migrates through the aquifer. In this approach it is useful to introduce the differential operator

$$\frac{D}{Dt} = \frac{\partial}{\partial t} + u_i \frac{\partial}{\partial x_i} \tag{10.72}$$

This is called the *material derivative* and represents the rate of change with respect to time of a quantity associated with a particle moving with velocity **u**. In other words, it shows how the changes in the quantity appear to an observer sitting on the moving particle. As an illustration, equation (10.31) with D equal to zero describes transport by advection only. If flow is uniform, the governing equation becomes

$$\frac{\partial}{\partial t}(nc) + u_i \frac{\partial}{\partial x_i}(nc) = 0 \tag{10.73}$$

The Lagrangian form of equation (10.73) is

$$\frac{D}{Dt}(nc) = 0 \tag{10.74}$$

This equation states that the rate of change of solute concentration associated with a fluid particle is zero, which is the expected result in purely advective flow. Of greater interest is the equation which describes transport by advection and dispersion in a non-deforming medium with steady flows:

$$\frac{\partial}{\partial t}(nc) + u_i \frac{\partial}{\partial x_i}(nc) = \frac{\partial}{\partial x_i}\left(D_{ij}\frac{\partial c}{\partial x_j}\right) \tag{10.75}$$

The Lagrangian form of this equation is

$$\frac{D}{Dt}(nc) = \frac{\partial}{\partial x_i}\left(D_{ij}\frac{\partial c}{\partial x_j}\right) \tag{10.76}$$

and this describes how the solute concentration associated with a fluid particle changes with time due to dispersive effects. This formulation of the equation suggests that the solution of the problem might be tackled using a two-stage process. Assuming that the flow equation has been solved to produce a fluid velocity field, the strategy is to introduce theoretical water particles into the model domain. Each particle is assigned a value which represents the solute concentration at the position of the particle. At each time step:

(1) the positions of the particles are updated by moving them according to the velocity field, i.e. with the flow of water through the medium, and

(2) the concentrations associated with the particles are then updated according to equation (10.76).

The steps of the solution can be expressed more formally as:

(1) Let $x_p(t)$ be the position of particle p at time t, and write $x_p^m = x_p(m\Delta t)$. Let u_p be the velocity of particle p. Then x_p^{m+1}, the position of particle p at time $(m + 1)\Delta t$, can be determined by solving the initial-value problem

$$\frac{dx_p}{dt} = u_p \tag{10.77}$$

$$x_p(m\Delta t) = x_p^m$$

(2) Let $c_p(x_p, t)$ be the concentration associated with particle p at x_p at time t, and let $c_p^m = c_p(x_p^m, m\Delta t)$. Then c_p^{m+1}, the concentration associated with particle p at time $(m + 1)\Delta t$, is determined by solving the system of initial value problems

$$\frac{\partial}{\partial t}(nc_p) = \frac{\partial}{\partial x_i}\left(D_{ij}\frac{\partial c_p}{\partial x_j}\right) \tag{10.78}$$

$$c_p(x_p^m, m\Delta t) = c_p^m$$

(where the index p ranges from 1 to the total number of particles) subject to the concentration boundary conditions of the problem.

A number of techniques for solving the system of equations in step 2 have been developed. Some methods employ a fixed grid in which the value at each node represents the concentration at the location of the node. In this case, nodal values are estimated by interpolation from the particle concentrations and are then used to produce estimates of the spatial derivatives used in step 2. The form of the interpolator used can have a significant effect on the accuracy of the method. Other approaches involve the estimation of the spatial derivatives from the point values themselves.

These techniques for solving highly advective transport problems, which are often referred to as *moving-point* methods, produce very accurate results in homogeneous media. However, the computational effort involved in tracking a large number of particles can be high, and a number of schemes have been developed in which large numbers of particles are used only where they are required, namely in the region of sharp concentration fronts (Neuman 1984; Farmer 1985). Yeh (1990) tackles the problem of computational expense by using a method of adaptive mesh refinement. The implementation of these adaptive moving-point techniques requires the construction of more complex computer programs, but the quality of the results fully justifies the programming effort involved. The further development of moving-point methods is an interesting area of research.

10.5 SAMPLE COMPUTER PROGRAM

The FORTRAN program below represents a one-dimensional model for simulating
the migration of a tracer in groundwater. A finite difference grid is superimposed on
the modelled domain and tracer is considered to be injected at the node of the first
grid element. A sample of the input file, CONTROL.DAT, is shown below. The initial
contaminant distribution is considered to be uniform and is referred to in the input file
as background concentration. The 'inflow' boundary condition can be set to simulate
continuous or pulse injection of tracer. This is done by setting the times at which
tracer injection begins and ends. A fixed concentration boundary condition is imposed
at the 'outflow' end of the domain. This is set automatically to the background
concentration. The influence of the fixed concentration boundary condition can be
significant, and it is recommended, therefore, that the mesh data are defined so that
only small amounts of contaminant reach the outflow boundary. The program solves
equation (10.68), subject to the above initial and boundary conditions, to produce the
concentration of tracer at each node in the grid at a time defined by the product of the
length of each time step and the number of time steps. These concentration values are
written to the output file, MODEL.OUT.

The solution technique can be chosen from the explicit, fully implicit or Crank–
Nicholson schemes by setting the appropriate value in the input file. It is suggested
that the reader uses this model to become familiar with the relative merits of each
solution technique.

```
***********************************************************************
*  This program solves the advection–dispersion equation describing
*  steady state flow from a point contaminant source using finite
*  difference methods.
***********************************************************************

       implicit real (a − h, l − z)

       real m(50,4), c(50)

***********************************************************************
*  Read CONTROL.DAT
***********************************************************************

       open (unit = 10, file = 'CONTROL.DAT')
       read (10,*)
       read (10,*)
       read (10,*) indx
       read (10,*)
       read (10,*) dx
```

```
        read (10,*)
        read (10,*) indt
        read (10,*)
        read (10,*) dt
        read (10,*)
        read (10,*)
        read (10,*)
        read (10,*) pump1
        read (10,*)
        read (10,*) pump2
        read (10,*)
        read (10,*) bgrnd
        read (10,*)
        read (10,*) concin
        read (10,*)
        read (10,*)
        read (10,*) u
        read (10,*)
        read (10,*) aL
        read (10,*)
        read (10,*) ne
        read (10,*)
        read (10,*)
        read (10,*) theta
        close (10)

************************************************************************
* Open output file MODEL.OUT
************************************************************************

    open (unit = 11, file = 'MODEL.OUT')

************************************************************************
* Set initial concentrations
************************************************************************

    do 100 i = 1, indx
      c(i) = bgrnd
100 continue

************************************************************************
* MAIN PROGRAM LOOP
************************************************************************
```

```
      do 300 istep = 1, indt
*********************************************************************
* Set concentration of injection node
*********************************************************************

      tim = float(istep)*dt
      if (tim .lt. pump1 .or. tim .gt. pump2) then
        c(1) = bgrnd
      else
        c(1) = concin
      end if


*********************************************************************
* Set up matrix
*********************************************************************

      Pe = ne*dx/aL
      do 200 i = 1, indx − 2
      m(i,1) = − theta*(1.0 + Pe/2.0)
      m(i,2) = 2.0*theta + dx**2/(al*u*dt/ne)
      m(i,3) = − theta*(1.0 − Pe/2.0)
      m(i,4) = (1.0 − theta)*(1.0 + Pe/2.0)*c(i)
   $          − (2.0*(1.0 − theta) − dx**2/(al*u*dt/ne))*c(i + 1)
   $          + (1.0 − theta)*(1.0 − Pe/2.0)*c(i + 2)
200  continue


*********************************************************************
* Impose boundary conditions
*********************************************************************

      m(1,4) = m(1,4) + theta*(1.0 + Pe/2.0)*c(1)
      c(indx) = bgrnd
      m(indx − 2,4) = m(indx − 2,4) + theta*(1.0 − Pe/2.0)*c(indx)


*********************************************************************
* Solve matrix equation
*********************************************************************

      call tridiag (indx − 2, m, c)
300  continue


*********************************************************************
* Write results to file MODEL.OUT
*********************************************************************
```

```
      do 400 i = 1, indx
         write (11,*) c(i)
400 continue
      close (11)
      stop
      end


**********************************************************************
* Subroutine tridiag
* This subroutine solves the system of linear equations
**********************************************************************

      subroutine tridiag (n, m, c)

      real m(50,4), c(50)
      integer n

      do 1000 i = 2, n
         m(i,2) = m(i,2) - (m(i,1)/m(i - 1,2))*m(i - 1,3)
         m(i,4) = m(i,4) - (m(i,1)/m(i - 1,2))*m(i - 1,4)
1000 continue
      m(n,4) = m(n,4)/m(n,2)
      do 2000 i = n - 1,1, - 1
         m(i,4) = (m(i,4) - m(i,3)*m(i + 1,4))/m(i,2)
         c(i + 1) = m(i,4)
2000 continue
      return
      end
```

CONTROL.DAT

MESH DATA
Number of grid elements
   50
Size of each grid element (metres)
   0.1
Number of time steps
   16
Length of each time step (days)
   0.1

BOUNDARY AND INITIAL CONDITIONS
Time at which injection begins (days)
   0.0
Time at which injection stops (days)
   0.5
Background contaminant concentration
   0.0
Injected contaminant concentration
   1.0

TRANSPORT DATA
Groundwater velocity (metres/day)
   1.0
Dispersivity (metres)
   0.1

SOLUTION TECHNIQUE
(explicit: 0, Crank-Nicholson: 0.5, fully implicit: 1)
   0.5

REFERENCES

Bear, J. (1972) *Dynamics of Fluids in Porous Media*. Elsevier, New York.

Farmer, C. L. (1985) Moving point techniques. In *Advances in Transport Phenomena in Porous Media*, Eds. Bear, J. & Corapcioglu, M. Y., NATO ASI Series, Martinus Nijhoff, Dordrecht.

Freeze, R. A. and Cherry, J. A. (1979) *Groundwater*. Prentice-Hall, Englewood Cliffs, New Jersey.

Huyakorn, P. S. and Pinder, G. F. (1983) *Computational Methods in Subsurface Flow*. Academic Press, New York.

Neuman, S. P. (1984) Adaptive Eulerian–Lagrangian finite element method for advection dispersion. *International Journal of Numerical Methods in Engineering*, **20**, 321–37.

Prickett, T. A., Naymik, T. G. and Lonnquist, C. G. (1981) A 'random walk' solute transport model for selected groundwater quality evaluations. Bulletin 65, Illinois State Water Survey, Champaign, Illinois.

Yeh, G. T. (1990) A Lagrangian–Eulerian method with zoomable hidden fine-mesh approach to solving advection–dispersion equations. *Water Resources Research*, **26**(6), 1133–44.

Chapter 11

Modelling of Sewage Outfalls in the Marine Environment

S. MARKHAM

11.1 INTRODUCTION

Many coastal communities dispose of their domestic and industrial waste to the sea through an outfall. A marine outfall system must ensure compliance with the environmental quality objectives (EQOs). These EQOs require that a body of water should be suitable for its designated purpose such as bathing or shellfish and these uses are protected by one or more environmental quality standards (EQSs). In the UK bathing and shellfish waters are covered by the EEC directives (76/160/EEC) and (79/923/EEC) respectively. These directives are designed to improve and protect the environment and are useful to the engineer because they provide quantitative parameters on which to assess an outfall design.

To keep disposal to the marine environment in perspective, a useful maxim is that the oceans have a capacity for near unlimited dilution but for conservative substances there is a near infinite retention time.

The term 'marine disposal' is misleading when applied to sewage, as most components of domestic sewage are readily broken down in sea water. There is usually an abundant supply of oxygen; this stimulates oxidation and the decomposed material then returns into the lower orders of the food web. The term 'marine treatment' is an improved description.

Sewage contamination of seawater has been shown to be a health risk for bathers (Cabelli 1979). The elements of sewage that are hazardous to humans are the pathogenic bacteria and viruses. Monitoring specific pathogens is at present considered impractical; however, pathogenic microorganisms present in sewage are usually

An Introduction to Water Quality Modelling. 2nd Edition. Edited by A. James
© 1993 John Wiley & Sons Ltd.

outnumbered by commensal bacterial flora of the human intestine (i.e. *E. coli*). The EEC directives (76/160/EEC) and (79/923/EEC) relies heavily on *E. coli* for the indication of pathogens in the sea environment.

The natural breakdown of sewage (including the mortality of microorganisms) is concurrent with the physical mixing of the effluent and ambient sea water. These two phenomena allow the receiving water to meet the relative EQSs.

11.2 PHYSICAL PROCESSES

After sewage effluent is discharged from an outfall the mixing process can be characterized in two stages.

The first stage of sewage discharged into the sea is where the effluent moves upwards due to the bouyant force created by the density difference between sea water and sewage. Dilution occurs as the buoyant jet entrains sea water at its perimeter. An understanding of the mechanisms involved when sewage flows into the sea can be developed from the theory of jets and plumes. This is covered in more detail in Fischer *et al.* (1979); a descriptive summary follows.

A jet is the discharge of fluid from an orifice into a large body of the same or similar fluid; here the flow is driven by the momentum of the discharge. A plume is a flow caused by the potential energy of the fluid given by its positive or negative buoyancy relative to its surroundings. Discharges of sewage into the marine environment are classed as buoyant jets, and are derived from sources of both momentum and buoyancy. Sewage leaves the outfall in a similar fashion to a jet with essentially a uniform velocity profile. At the jet boundary turbulent shear develops and this gradually spreads towards the centre line of the jet. The velocity flux is assumed to be constant with no entrainment of the ambient occurring. The length of this constant velocity flux was found to be 5-7 times the diameter of the exit port (Albertson *et al.* 1950). This region is called the *zone of flow establishment*. When the turbulence has reached the jet centre line the regime changes to the *zone of established flow*, the jet continues to expand and the mean velocities and sewage concentrations decrease until it reaches the surface in a diluted form—an *initial dilution*. If at the surface the *initial dilution* is greater than 1:50 the buoyancy driven force can then be neglected (Abraham and Brolsma 1965).

At the surface the sewage field is subjected to further mixing from the sea. This process is the second stage of mixing and is called here *subsequent dilution*. This *subsequent dilution* can be described, on the scale used in outfall models, as having an advective component (transport away from the outfall mouth) and a diffusive component (local mixing of the plume). It can be reasoned that this advection and dispersion is essentially the same process; it is just mixing over different ranges of scale where the large eddy advects and the smaller ones disperse. As an illustration consider the action of three different eddies on a surface plume shown in Figure 11.1.

(1) Eddies much larger than the plume element only cause advection and meandering.

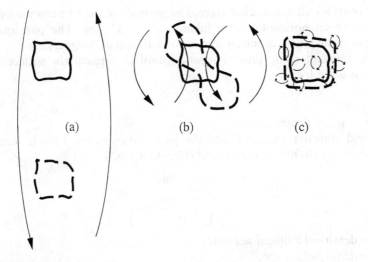

(a)   (b)   (c)

Figure 11.1. (a) eddy size $\gg$ plume dimension; (b) eddy size = plume dimension; (c) eddy size $\ll$ plume dimension

(2) Eddies approximating the plume element size cause dispersion. Here the plume element becomes distorted by localized velocity variations.

(3) The small turbulent eddies (small scale random velocity fluctuations) also facilitate the spreading of the plume by increasing the area over which molecular diffusion can take place.

As the surface plume grows the larger scale eddies, which first advected the plume, gradually become active in the mixing process.

11.3 HYDRAULIC DESIGN OF OUTFALLS

The velocity in the pipe should be high enough to maintain scour and prevent deposition of particles. This can be achieved if the velocities of 0.76 m/s for raw sewage or 0.5 m/s for settled sewage is reached once per day (Neville-Jones and Dorling 1986).

The outlet of an outfall may consist of a simple open end or multiport diffuser. The purpose of a multiport diffuser is to achieve a greater *initial dilution*. This is achieved because more effluent will entrain sea water between the outlet ports and the sea surface. A multiport diffuser provides an increased initial dilution within a small mixing zone. Nevertheless by the time diluted effluent has been advected a few diffuser lengths the plume dilution becomes independent of the diffuser length. Therefore a diffuser is really necessary only if the initial dilution cannot be achieved by an open ended pipe.

The use of a diffuser has been known to cause operational problems (Grace 1985). If a diffuser is selected its hydraulic design should ensure a near uniform division of flow between ports for all design discharges. The ports should not be less than 100 mm in diameter and set horizontal for a higher surface dilution. The port spacing is a function of rising plume geometry and in the UK outfall designers use a spacing that keeps the buoyant rising plumes separate until they reach the surface. The port spacing l is given by:

$$l = \tfrac{1}{3} H \tag{11.1}$$

where H = water depth.

To avoid seawater intrusion into the pipe a densimetric Froude number (F_D) greater than one should be maintained (Charlton 1985).

$$F_D = \frac{u_j}{\left[\dfrac{(\rho_a - \rho_j)}{\rho_j} gD \right]^{1/2}} \tag{11.2}$$

ρ_a = density of ambient sea water;
ρ_j = density of sewage;
g = acceleration due to gravity.

In an effective design the diffuser pipe steps down in size to maintain scour velocities and Froude numbers greater than unity.

11.4 THE MODEL OUTLINE

This outfall model is essentially a combination of three models. The first part calculates the initial dilution S_a, the second part the subsequent dilution S_d, and the third part bacterial mortality S_b (represented as dilution). The concentration of E. coli in the outfall pipe is in the order of $10^7/100$ ml and it has become a custom in the UK to design outfalls to the mandatory standard for E. coli in the bathing water directive (76/160/EEC) of 2000 E. coli/100 ml. Therefore a total dilution, S_t, greater than 5000 is required; however, a large safety margin, perhaps a factor of 4, is advisable even if the parameters used have been measured in situ.

11.5 INITIAL DILUTION

A good estimate of initial dilution is the first stage of the outfall model. For calm, homogeneous receiving waters this is not particularly difficult and can be reliably predicted using mathematical models (Fan and Brooks 1969). In moving water there appears no universally accepted model of prediction; mathematical models have been verified only against laboratory data and then for momentum-dominated discharges (high densimetric Froude numbers) in low ambient currents, mainly for vertical buoyant jets. Sea outfalls in the UK have the characteristics of horizontal jets and low densimetric Froude numbers and rarely deal with discharge into a stationary sea.

Around British waters designers tend to use empirical correlations developed from field observations (Agg and Wakeford 1972; Bennett 1983) where the dimensionless parameters F_D, H/D and u_a/u_j are used to characterize the problem. However recent research (Lee and Neville-Jones 1987) has produced a method of predicting initial dilution founded on the fundamental fluid flow concepts of volume flux Q, momentum flux M and buoyancy flux B; for more detail see Fischer *et al.* (1979).

Volume flux:

$$Q = \frac{u_j \pi D^2}{4} \tag{11.3}$$

where D = outfall pipe or diffuser port diameter;
$\quad\; u_j$ = velocity from outfall pipe or diffuser port.

Discharge specific buoyancy flux:

$$B = Q \left[\frac{\rho_a - \rho_j}{\rho_a} \right] g \tag{11.4}$$

Buoyancy-dominated near field (BDNF):

$$S_0 = 0.31 \frac{B^{1/3} H^{5/3}}{Q} \qquad Y < \frac{5B}{u_a^3} \tag{11.5}$$

Buoyancy-dominated far field (BDFF):

$$S_0 = 0.32 \frac{u_a H^2}{Q} \qquad Y \ge \frac{5B}{u_a^3} \tag{11.6}$$

u_a = ambient current velocity

S_0 is the minimum surface dilution. To select whether to use BDNF or BDFF a buoyancy length scale,

$$l_b = \frac{B}{u_a^3} \tag{11.7}$$

is used to measure the vertical distance within which buoyancy is important. BDNF and BDFF refer to plume-dominated and current-dominated regimes respectively.

To ensure that the immediate area of discharge is not polluted by smell, sewage slicks and solids it is necessary to achieve some minimum initial dilution at the sea surface. Sewage diluted by clean sea water is barely detectable to the eye at a concentration of less than 1:100 (Newton 1972) and this aesthetic consideration is the design objective in respect to initial dilution.

11.6   DETERMINATION OF PLUME SIZE

The size of the plume is required to calculate the dispersion coefficient K. If one considers a discharge from an open-ended pipe, a derivation of the plume geometry from Rawn and Palmer (1930) can be shown:

A buoyant jet from an outfall takes a nearly parabolic path to the surface, and the length of the path, L_p, is given by

$$L_p = 1.33 \frac{H}{D} \tag{11.8}$$

The thickness of the pollutant field, H_t, was then show to be given by

$$H_t = \frac{L_p}{12} \tag{11.9}$$

If one assumes that the surface distribution of the pollutant is radially symmetric, then consider a release over a time interval dt. The volume of material released is therefore $= Q\, dt$. If the initial dilution is $= S_0$, then the volume V_m of material in the patch is given by

$$V_m = S_0 Q\, dt \tag{11.10}$$

If we say the standard deviation of the distribution is σ, then the volume V_m of the patch defined by a radius of a standard deviation is therefore

$$V_m = 2\pi\sigma H_t \tag{11.11}$$

$$2\pi\sigma H_t = S_0 Q\, dt \tag{11.12}$$

and

$$\sigma = \frac{S_0 Q\, dt}{2\pi H_t} \tag{11.13}$$

If the initial size of the patch is assumed to be radius R_0

$$R_0 = 2\sigma \tag{11.14}$$

(this covers 97.7% of the material present), then

$$R_0 = 2\sigma = \frac{2 S_0 Q\, dt}{\pi H t} \tag{11.15}$$

To calculate the plume width from a diffuser compute the size of the plume from the first and last port. Then add half the plume width from the first and last port to the length of the diffuser, the diffuser length can be calculated from equation (11.1). The length scale used to calculate the diffusion coefficient K also depends on the angle at which the sea current crosses the diffuser: the plume width will be a minimum when parallel to the diffuser and a maximum when perpendicular.

11.7 COASTAL CURRENT

Of all the physical oceanographic features about which data are required, currents are the most complex. When the times of low tide differ along a coast, i.e. when the co-tidal lines are not parallel to a coast, a tidal current flows alongshore. Being associated with the tidal wave current it reverses its direction during a tidal cycle. The tidal current is caused by the gravitational pull of the Sun and the Moon when the Earth rotates. Together with the tidal streams there are those currents generated by the wind. A summary of individual investigations of wind-induced currents show that the induced current is in the same direction as the wind and has a speed of about 3% of the wind speed, measured at 10 m above the Earth's surface (Bowden 1984). On the other hand using Ekman's theory it can be shown that in the deep open ocean the wind-induced surface current is deflected 45° to the right of the wind direction (in the northern hemisphere and left in the southern hemisphere), and with greater depths the deflection increases while the velocity decreases. The underlying mechanisms involve the Coriolis force and the vertical eddy viscosity (Pond and Pickard 1978).

In shallower water near the coast the current profile is modified. Bottom friction provides an additional shearing force. The deflection by the Coriolis force diminishes. A sea surface slope is produced at the coastal boundary and this gives rise to a secondary current which tends to be uniform with depth.

A difference develops between the current velocity profile generated by winds blowing perpendicular to the shore and winds parallel to the shore. The relative velocities produced are best shown with the mathematical model analysis by Gauthier and Quentin (1977).

Winds normal to the coast are shown in Figure 11.2: the calculated current profile is for when the wind is directed towards the shore. The velocities fall off very quickly with increase in depth and become zero at about 40% of the depth below which a compensatory current flows off shore. The mean shoreward transport velocity of an effluent would depend on the percentage of the water column occupied by sewage (see equation (11.9)). There should be some reservations regarding the high velocity predicted at the sea surface especially when the wave height is high.

For winds parallel to the coast the current profiles are a different shape, as shown by Figure 11.3. For the fully developed current system the velocities are greater than for onshore winds. There is a rapid fall of velocity with increase in depth and again there must be some reservation about the high surface velocities.

From Figures 11.2 and 11.3 the Coriolis force increases the wind induced velocity and the question whether the Coriolis force is a factor in a shallow coastal water can be assessed using the Ekman number E_{num}

$$E_{num} = 0.3H^{0.45}W^{-0.50} \qquad (11.16)$$

where H = water depth;
$\quad\quad W$ = wind speed.
(This applies to a latitude of 45°.)

The Coriolis force is said to be significant when this number is > 1 and negligible when < 0.3.

It also takes time (t) for a wind-induced current to become established and Gauthier (1976) suggested the relationship:

$$t \simeq \frac{11610\,H^{7/6}}{W} \qquad (11.17)$$

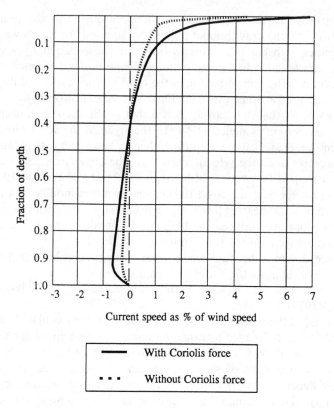

Figure 11.2. Winds normal to the coast

where t = time (s);
H = water depth (m);
W = wind speed (m/s).

11.8 CALCULATING *SUBSEQUENT DILUTIONS* USING BROOKS' (1960) METHOD

The model was developed to solve the centre-line concentration (or *subsequent dilution*) in a surface plume transported in a uniform current.

11.8.1 Assumptions

(a) The current has a constant and uniform speed.
(b) Vertical mixing is negligible.
(c) Mixing in the direction of the current (x) is negligible.

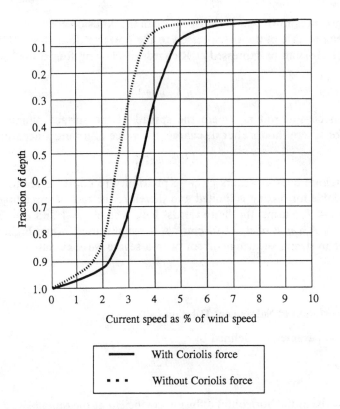

Figure 11.3. Winds parallel to the coast

(d) Mixing in the transverse direction (y) is described by a diffusion process with a variable diffusion coefficient, K, which is a function of plume width.
(e) The surface plume moves with the surface current system.
(f) The reduction in coliform concentrations due to physical dilution and due to die-off are assumed to be independent.

11.8.2 Horizontal Diffusion Coefficient (K)

The key parameter in the determination of subsequent dilution by the method of Brooks is the horizontal diffusion coefficient (K). K is not a constant but increases at some power of a length scale (L).

$$K = \alpha L^n \tag{11.18}$$

where L = length scale (this is taken as the surface plume width perpendicular to the average current direction).
α = a dissipation parameter.
n = a power exponent.

Experience has shown that in the open ocean the diffusion coefficient is proportional to the length scale of the patch raised to the power of four thirds (Okubo, 1974). Equation (11.18) can be expressed as Richardson's law or 'four thirds' law (Stommel 1949).

$$K = \alpha L^{4/3} \qquad (11.19)$$

For open coastal waters, where the spread of the surface plume can proceed unbound for several hours after discharge, the value of n ranges according to

$$1.0 \leq n \leq \tfrac{4}{3}$$

For nearshore coastal waters in close proximity to coastline features the size of eddies available for mixing is limited and therefore the rate of spreading is reduced.

To estimate K_0 using the 'four thirds' law, where $n = \tfrac{4}{3}$ and $\alpha = 0.02$, 0.01 or $0.005 \text{ cm}^{2/3}/\text{s}$. A value of $\alpha = 0.01 \text{ cm}^{2/3}/\text{s}$ is usually used, while higher values are appropriate in strong shearing current or in a severe wave climate.

11.8.3 Calculation of Subsequent Dilution

Calculate the parameter β defined by

$$\beta = \frac{12K_0}{Ub} \qquad (11.20)$$

where K_0 = the initial horizontal diffusion coefficient at the outfall site.

b = surface plume width over the outfall pipe or diffuser perpendicular to the current.

Calculate the parameter P

$$P = \frac{\beta x}{b} \qquad (11.21)$$

where x = the distance along the current path to the site where dilutions (or concentration) is required.

Dimensionless field width:

$$\frac{L_x}{b} = (1 + (2 - n)P)^{[1/(2-n)]} \qquad (11.22)$$

where L_x is the width of the plume at distance x and n is the same value used to calculate K_0.

Minimum subsequent dilution:

$$S_d = \frac{C_0}{C_{max}} = \frac{1}{\text{erf}\left[\dfrac{1.5}{(1 + (2-n)P)^{2/(2-n)} - 1}\right]^{1/2}} \qquad (11.23)$$

The error function is defined by the equation:

$$\text{erf}\,\{X\} = \left[\frac{1}{\sqrt{\pi}}\right]\int_0^X \exp(-p^2)\,dp \tag{11.24}$$

The error function can be calculated using Simpson's rule.

11.8.4 Microbial Mortality

Microbial mortality is a significant component in an outfall system achieving the required EQO. The mortality is measured as a T_{90}, which is the time in hours for the inactivation of 90% of the initial number of microbes. Studies of mortality rates have demonstrated light to be the most important single factor responsible for the inactivation of *E. coli* in sea water. Since light intensity varies with weather conditions and water transparency there will be a range of T_{90} values (Gould and Munro 1981). For clear coastal waters around Britain a T_{90} of 6–12 h is a reasonable design parameter. The component of microbial reduction can be expressed as a dilution S_b.

$$S_b = 10^{[t/T_{90}]} \tag{11.25}$$

where T_{90} = time (s) for microbial density to decrease 90%.
$\quad\quad t$ = travel time (s) to where dilution is measured.

11.9 CONCENTRATION AT TARGET AREA

The total dilution S_t at any target area is the product of the component dilutions.

$$S_t = S_0 S_d S_b \tag{11.26}$$

and for conservative substances $S_b = 1.0$.

The maximum concentration at a target area C_x can be calculated for any contaminant by using

$$S_t = \left[\frac{(C_e - C_b)}{(C_x - C_b)}\right] \tag{11.27}$$

where C_e = effluent concentration in outfall pipe;
$\quad\quad C_x$ = maximum concentration at the targert area;
$\quad\quad C_b$ = background concentration.

11.10 WORKED EXAMPLE

With the introduction of the friendly personal computer, at the beginning of the 1980s there came a selection of usable software packages. One type of such packages was the spreadsheet. There are many spreadsheet software packages available but the underlying method of use is the same for all. The outfall model outlined above can be programmed onto a spreadsheet as the following example shows.

A north-east town wishes to discharge to the marine environment a dry weather flow of 0.13 m³/s of raw sewage. An available outfall site is to the south of a designated bathing beach and the community can afford an outfall pipe length of 2 km. The low water tidal current runs north at 0.7 m/s and the wind rose shows a NE wind to be the most predominant during the summer months with a strength of 5 m/s.

Appendix A gives a listing of an outfall model written on a spreadsheet; the listing of the individual cell entries is also included. When running the model it is prudent to test its sensitivity to the parameters.

11.11 SOME TIPS ON SPREADSHEET CONSTRUCTION

Before writing into the spreadsheet do a sketch of the program on a sheet of paper showing the location for the data, main calculation and macros. When building a formula into a particular cell use the 'point', i.e. cursor, to move the highlight to the actual cell you are using in the calculation. Build the formula in stages to help eliminate error. 'Range Names' can be given to a cell or range of cells instead of using the letter and number reference; this gives a more readable model.

In designing the program make the spreadsheet cells do as much as possible of the calculation because it is more efficient than manipulation with macros and easier to, debug.

REFERENCES

Abraham, G. and Brolsma, A. A. (1965) *Diffusers for Disposal of Sewage in Shallow Tidal Water*. Delft Hydraulics Lab. Pub. No. 37.

Agg, A. R. and Wakeford, A. C. (1972) Field Studies of jet dilution of sewage outfalls. *Instn. Publ. Hlth. Engrs. J*, **71**, 126–49.

Albertson, M. L., Dia, Y. B., Jensen, R. A. and Rouse, H. (1950) Diffusion of submerged jets. *Trans. ASCE*, **115**, 639–97.

Bennett, N. J. (1983) Design of sea outfalls—the lower limit concept of initial dilution. *Proc. Instn. Civ. Engrs*, Part 2, **75**, 113–21.

Bowden, K. F. (1983) *Physical Oceanography of Coastal Waters*. Ellis Horwood, Chichester.

Brooks, N. H. (1960) Diffusion of sewage effluent in an ocean current. In *Waste Disposal in the Marine Environment*, Ed. E. A. Pearson, Pergamon Press, Oxford.

Cabelli, V. J., Dufour, A. P., Levin, M. A., McCabe, L. J. and Haberman, P. W. (1979) Relationship of microbial indicators to health effects at marine bathing beaches, *AJPH*, **69** (7).

Charlton, J. A. (1985) Sea Outfalls. In *Developments in Hydraulic Engineering—3*. Ed. P. Novak, Elsevier Applied Science, Barking.

Fan, L. N. and Brooks, N. H. (1969) *Numerical Solutions of Turbulent Buoyant Jet Problems*. W. M. Keck Laboratory of Hydraulics & Water Resources, California Institute of Technology, Pasadena, Report KHR-18.

Fischer, H. B., List, E. J., Koh, R. C. Y., Imberger, J. and Brooks N. H. (1979) *Mixing in Inland and Coastal Waters*. Academic Press, New York.

Gauthier, M. F. (1976) Modèles de calcul des écoulements induits par le vent. *XIV Journées de l'hydrauliques Société Hydrotechnique de France*, 1–9.

Gauthier, M. F. and Quentin, B. (1977) Modèles mathématiques de calcul des écoulements induits par le vent. *17th Congress of the International Association of Hydraulic Research* **3**, 69–76.

Gould, D. J. and Munro, D. (1981) Relevance of microbial mortality to outfall design. *Coastal Discharges*. Thomas Telford, London.

Grace, R. A. (1985) Sea outfalls—a review of failure damage and impairment mechanisms. *Proc. Instn. Civ. Engrs.* Part 1, **77**, 137–52.

Lee, J. H. W. and Neville-Jones, P. (1987) Sea outfall design-prediction of initial dilution. *Proc. Instn. Civ. Engrs.*, Part 1 **82**, 981–94.

Neville-Jones P. J. D. and Dorling, C. (1986) *Outfall Design Guild for Environmental Protection, a Discussion Document.* ER 209E, WRc Wiltshire.

Newton, J. R. (1975) Factors affecting slick formation at marine sewage outfalls. In *Pollution Criteria for Estuaries*, Ed. R. R. Helliwell & Bossayni, Pentech Press, London, pp. 12.1–12.6.

Okubo, A. (1974) Some speculations on oceanic diffusion diagrams. *Rapports et Procès verbaux des Réunions Conseil international pour L'Exploration de la Mer*, **167**, 77–85.

Pond, S. and Pickard, G. L. (1978) *Introductory Dynamic Oceanography*. Pergamon Press, Oxford.

Rawn, A. M. and Palmer, H. K. (1930) Predetermining the extent of sewage field in sea water. *Trans. Am. Soc. Civ. Engrs*, **94**, 1036–81.

Stommel, H. (1949) Horizontal diffusion due to oceanic turbulence. *Journal of Marine Research*, **8** (3), 199–225.

APPENDIX A: MODELLING OF SEWAGE OUTFALLS IN THE MARINE ENVIRONMENT

```
# # # # # # # # # # # # # # # # # # # # # # # # # # # # # # # # # #

#   Design Parameters   #

# # # # # # # # # # # # # # # # # # # # # # # # # # # # # # # # # #
```

Outfall Data

Discharge DWF (raw sewage)	0.13	m^3/s
Pipe diameter	0.65	m
Distance to model	2000	m
E. coli concentration in effluent	1.00E + 07	number
Density of sewage	1000	kg/m^3
Acceleration due to gravity	9.81	m^2/s

Marine Data

Water depth	22	m
Tidal current direction	200	?
Tidal current speed	0.05	m/s
Wind direction	135	?
Wind speed	4	m/s
Density of sea water	1025	kg/m^3
Factor *n*	1.1	number

Dispersion parameter   0.005   cm$^{2/3}$/s
Background *E. coli* concentration   20   number
T 90 for *E. coli*   8   h

###############

Outfall Model

###############

Hydraulic Analysis

(2*DWF is reached once per day)
Outfall pipe velocity   0.783532   m/s
                     There is adequate scour

Sea Current

(Assume the wind induces a current of 3% of the wind
speed and in the same direction as the wind)

North–south component   0.037868   m/s
East–west component   − 0.10195   m/s
Sea current direction   249.6238   ?
Sea current speed   0.108759   m/s

Initial Dilution

Specific Buoyancy Flux   0.031105
               Buoyancy dominated   near field
Initial dilution   129.5741

Subsequent Dilution

Plume geometry
Path length of plume   45.01538   m
Thickness of plume   3.751282   m
Plume size (for length scale)   571.7314   cm

Method of Brooks for calculating subsequent dilution

Horizontal diffusion coefficient (K)   0.005394   m^2/s
Dimensionless length β 0.104089
Dimensionless distance p 36.41179

Error function   0.024533

Simpson's Calculation of Error Function

Error	0.024533
Number of divisions	100
Step size	0.000245
New (x) ⟶	0.024533

First value of function ⟶	1
Accumulative value of P ⟶	50.61838
Accumulative value of Q ⟶	50.6308
Last value of function ⟶	1.024837

Value of Intergral ⟶	0.028214
Minimum dilution ⟶	35.44292

```
\a          {LET newx,0}
            {BLANK E89..E92}
            {LET firstvalue,@EXP(Newx)}
            {FOR counter, 1, division, 1, simpson}  counter ⟶
            {LET lastvalue,@EXP(newx)}
            {CALC}

simpson     {LET newx,newx + stepsize}
            {IF(@INT(counter) − 2*@INT(counter/2)) = 1}
            {LET pvalue,pvalue + @EXP(newx)}
            {IF(@INT(counter) − 2*@INT(counter/2)) = 0}
            {LET qvalue,qvalue + @EXP(newx)}
```

Field width of plume	49.93173	m
Subsequent dilution	35.44292	

Microbial Mortality

Time to reach target area	18389.24	s
Mortality expressed as a dilution	4.350259	

Concentration at Target Area

Total dilution	19978.5
E. coli concentration at target area	520.537

E41: 2*DISCHARGE/((PIPEDIAMETER/2)ʼ2*@PI)
E42: @IF(E41 < 0.76, 'Inadequate Scour', 'Adequate Scour')
E49: (@COS((CURRENTDIR*2*@PI)/360)*CURRENTSPEED) +
     (@COS(((WINDDIR + 180)*2*@PI)/360)*(WINDSPEED*0.03))

E50: (@SIN((CURRENTDIR*2*@PI)/360)*CURRENTSPEED) +
(@SIN(((WINDDIR + 180)*2*@PI)/360)*(WINDSPEED*0.03))
E51: @IF(E49 > 0#AND#E50 >
0,(@ATAN(E50/E49)*360)/(2*@PI),@IF(E49 < 0#AND#E50 > 0.90 +
(@ATAN(E49/E50)*360)/(2*@PI),@IF(E49 < 0#AND#E50 < 0.180 +
(@ATAN(E50/E49)*360)/(2*@PI),@IF(E49 > 0#AND#E50 < 0.270 +
(@ATAN(E49/E50)*360)/(2*@PI),0))))
E52: @SQRT(E49^2 + E50^2)
E56: +
DISCHARGE*((SEADENSITY − SEWAGEDENSITY)/SEADENSITY)*GRAVITY
E57: @IF(WATERDEPTH <
(5*DISCHARGE*((SEADENSITY − SEWAGEDENSITY)/SEADENSITY)*GRAVITY)
/(CURRENTSPEED^3), 'Near Field', 'Far Field')
E59:   @IF(E57 = 'Far   Field',   (0.31*(E56^0.33333)*WATERDEPTH)/DISCHARGE
(0.32*E52*(WATERDEPTH^2))/DISCHARGE)
E66: 1.33*WATERDEPTH/PIPEDIAMETER
E67: + E66/12
E68: (2*(2*E59*DISCHARGE*1)/(@PI*E67))*100
E73: (+ DISP*(E68^FACTORN))/1000
E74: 12*E73/(E52*(E68/100))
E75: + E74*DISTANCE/(E68/100)
E78: (1.5/((1 + (2 − FACTORN)*E75)^(2/(2-FACTORN)) − 1))^0.5
E84: + E78
E85: 100
E86: + E84/DIVISION
E87: 0.024533308741994
E89: 1
E90: 50.618378818879
E91: 50.630798705477
E92: 1.0248367265608
E94: (2/(@SQRT(@PI)))*(STEPSIZE/3)*(FIRSTVALUE + (PVALUE*4) +
(QVALUE*2) + LASTVALUE)
E95: 1/E94
E110: (1 + (2 − FACTORN)*E75)^(1/(2 − FACTORN))
E111: + E95
E118: + DISTANCE/E52
E120: 10^(E118/(T90*3600))
E127: + E59*E111*E120
E129: ((ECOLISEWAGE − ECOLIBACK)/E127) + ECOLIBACK

Index